Economics for the Built Environment

Economics for the Built Environment

Marcus Warren BA(Hons) MA

Senior Lecturer in Urban Land Economics at De Montfort University

Butterworth-Heinemann Ltd
Linacre House, Jordan Hill, Oxford OX2 8DP

 PART OF REED INTERNATIONAL BOOKS

OXFORD LONDON BOSTON
MUNICH NEW DELHI SINGAPORE SYDNEY
TOKYO TORONTO WELLINGTON

First published 1993

British Library Cataloguing in Publication Data

Warren, Marcus
Economics for the Built Environment
I. Title
690

ISBN 0 7506 1603 2

Library of Congress Cataloguing in Publication Data

Warren, Marcus.
Economics for the built environment/Marcus Warren.
p. cm.
Includes index.
ISBN 0 7506 1603 2
1. Real estate development. 2. Construction industry. 3. Land use. I. Title
HD1390.W37 1993
338.4'7624–dc20 92–45691
CIP

Set by Hope Services (Abingdon) Ltd
Printed and bound in Great Britain

Contents

Introduction vii

Hints on how to use the book x

The economist's approach xi

Part One: 'Market analysis' and its application in the built environment

1 Introductory market theory 3
2 Developing market analysis for further, and more precise, application 29
3 Land and construction related case studies 45
3A The housing market 45
3B The market for land (with specific reference to building land) 52
3C The market for construction components 57
3D The market for building labour 60
4 Government intervention in land and construction markets 66
Conclusion 82

Part Two: The 'theory of the firm' in the built environment

5 The theory of costs and revenues, and its application to construction firms in the private sector in influencing their output decisions 85
6 The long-run planning decision: the theory of long-run costs 97
7 Industrial structure and the construction industry (the theory of the firm) 103
8 The construction industry 122
Conclusion 139

Part Three: The macroeconomy and the built environment

9 The economy: a simple macroeconomic model 143
10 Primary economic objectives and the construction industry 157
11 National and local government economic policy and the built environment 162
Conclusion 186

Part Four: Urban economic policy and the making of the built environment

12 The creation of the built environment 189
13 Private sector investment appraisal techniques and the urban environment 206
14 Public sector investment appraisal and the built envirionment 226
15 The rationale for public intervention in the built environment 234
Conclusion 254

Suggested questions and tasks 255

Index 261

Introduction

This book is primarily concerned with the *application* of *economic theory* to a wide range of 'real world' issues concerning modern real estate. As such it is designed to be of use to:

1 First- and second-year students of degree and diploma courses specifically tailored to the requirements of gaining professional entry into the world of construction and property. Examples of such courses are those of Building Surveying, Land (Estate) Management, and Construction, to name just a few.

2 Students examining issues of the built environment as a subsidiary subject on other courses. For example many Economics degrees, or degrees in Combined Studies, have third-year options looking at general urban matters, or more specific topics such as Housing or Transport.

3 Many courses that are not related to the built environment may find this a useful text simply because of its high degree of application. The level of application should enable the reader to quickly see the relevance and flexibility of the subject of economics for the study of any industry whether it be in the world of property, or the textile industry for example. In fact, after looking at the applications suggested in this book it would be an interesting exercise to see how you can adapt the ideas presented to analyse any other industry of your choice.

4 Practising professionals. Hopefully this text will demonstrate that theory can be successfully used to examine real-world issues, and would help people in practice understand more clearly why markets behave the way they do, or appreciate the reason for the structure of their industry, just to name two potential applications.

It is important to realize at the outset however, that the book has been developed as a *companion*, albeit a very comprehensive one, to additional information that is gathered on courses via formal lecture, studio, and tutorial contact. Furthermore, and most importantly, when specializing in certain areas, or pursuing a personal interest within this book, it would be advisable to seek out more detailed theoretical information from one of the many 'mainstream' economics texts that are available. Moreover, to further the debate on some of the applied issues, such as the debate on land, for example, it would be best to examine additional readings specific to that topic that will obviously expand upon the foundations that are laid out in the confines of this introductory text.

As *application* is felt to be the most important theme, this work will continuously utilize examples and case studies drawn from construction and the built environment. This approach should enable the reader to quickly see the great usefulness and scope that a basic understanding of economics can give them in their future academic and professional lives.

Although the basic framework of theory is set out in this book, and numerous ideas proposed with respect to its application, it is left to the reader to search out up-to-date, relevant *data* so as to test and examine the theories in more detail. This enables one to apply this text over time, and use it in different countries with different data. Importantly though, it ensures that you become aware of the great range of data sources that are now available in this field of land, property, and construction.

Hopefully this work will also go some way to rectify the fact that experience has shown that, despite the increasing complexity of real estate issues in modern times, the subject of economics is often looked upon as being of secondary importance in construction or built environment courses. Typical claims made by professionals, lecturers, and students alike are: 'It is not as mainstream as "structures", "valuations", or "building technology", for example, as it is not directly related to buildings.' Or, even worse, some view it as being a complete irrelevancy. Sadly, such people are not only showing their ignorance about the broader educational process, but they also demonstrate their inability to envisage a subject such as building maintenance, for example, out of extremely narrow and unnecessary confines.

However, *naïveté* can, of course, be overcome, and I feel that once one has even a fundamental understanding of the main principles of economics one can quickly realize that it is not only a most diverse academic subject, but also in applied terms *everything* can be explained using some form of economic analysis or procedure. To illustrate my point, it can easily be demonstrated that the type of buildings that we see and build, the technology and processes that we use to build them, the materials that we use, the clients who occupy the completed building, and so on, are *all*, to some extent, governed by, or are reactions to, economic decisions and/or the impact of economic activity. Moreover, with respect to educational goals, students obviously need to appreciate the inter-linkages with all the other relevant subject areas. Without a thorough grounding of economics one would be at risk of educating students as technicians rather than as professionals who can rapidly analyse circumstances and forecast change, and are therefore in a situation to lead the market rather than merely react to it. Thus, such attitudes, if allowed to prevail, could hold back the progression of professionals who would be relying upon a technical rather than an analytical base.

Finally, it must be said that some perceive economics itself to be confined by its present 'neoclassical' barriers. However, my response is a similar one to that of the above concerning the keeping of any views in narrow confines. It should be possible to utilize basic economic principles and methods to see the advantages of such thought whether they be from the classical or neoclassical schools for example. By no means should one be so myopic as to view ideas without ongoing critical analysis, or to stay within the artificial boundaries of one school of thought. Such a standpoint would be as naïve as thinking that either capitalism or communism, in their purest forms, are the only way to run modern society.

Acknowledgements

My thanks to Dr John Ebohon for his advice and support, and to my wife, Jacqui, for drawing all the diagrams.

Hints on how to use this book

This text is written in a format that will hopefully make it a 'straightforward' read to the non-economist. However, in order to ensure that you have *understood* and *learnt* the principles that are introduced within the book you may find it useful to follow the recommendations set out below:

1 Firstly read the chapters as a whole at your normal reading speed This should enable you to get the general feel of the argument, and introduce you to any new terminology. Then, re-read them step by step so as to understand their fuller logic and content.

2 Make brief notes as you re-read the text.

3 Re-draw any diagrams and try to understand how they work and how they can be adapted to show different sets of circumstances.

4 Do not worry if you have to spend long periods of time understanding any one particular point or section. However, never go on without understanding the preceding information. Importantly, understand what you read rather than trying to memorize it.

5 Make a 'glossary' of technical terms. This can be useful as a point of reference, but the exercise will also help you to understand and remember key terminology.

The book is laid out in a series of parts. Each part covers a specific area and could be read in isolation from other parts. Nonetheless, there is a logical progression of ideas from beginning to end, and as such the maximum benefit would be derived from reading the text in its entirety. Within each part the text is divided into chapters, each chapter dealing with a specific sub-issue of the part. The design of this text is a response to much research indicating common student criticism concerning the somewhat arbitrary divisions of subjects between chapters that occur in some books.

The economist's approach

This brief section of the book is designed to give an *insight* into the methodology behind the subject of economics. It is by no means a comprehensive debate on scientific thought and processes; however, by appreciating a basic grounding of the methodology of the subject, common misunderstandings will hopefully be eradicated before one tackles the rest of the book.

Essentially one should perceive the subject of economics as a *science*, as scientific method enables us to enquire and analyse with a *logical* mind. Moreover, unless one understands this as the approach of the economist the subject matter of economics can be hard to follow. A guideline for the general methodology of economics follows below, and this information is summarized in Fig. 0.1.

Firstly, we need to make some *observations* about the issue that we wish to study (say, for example, the housing market) so that one can begin to formulate ideas and see what useful research can be undertaken. At this stage we

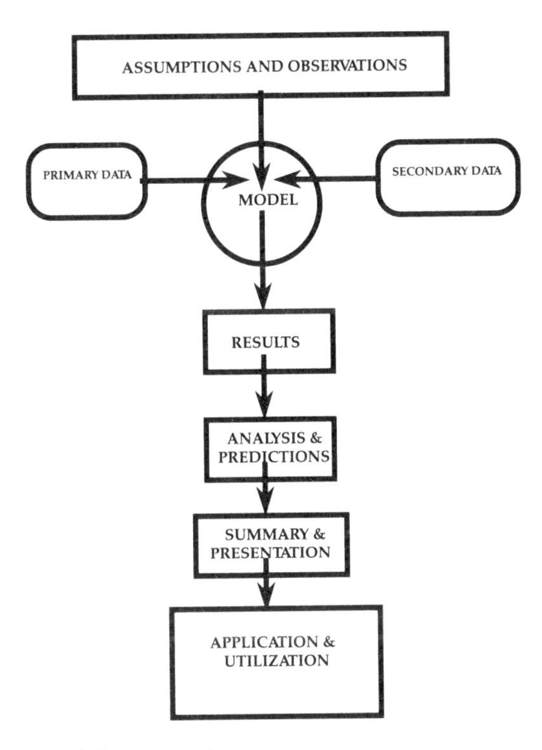

Figure 0.1. *The economist's approach*

may find that there are so many variables, or issues, to examine that in order to simplify our research, and most importantly so as to make it usable, we may have to introduce some *assumptions* into our analysis. Such assumptions can exclude information that is not of great importance, or can assume that other, more consistent, variables are constants. This simplifying procedure may make the model marginally less accurate but it may well be the only way to make it usable. As long as the end user is aware of the initial assumptions made these could well be changed, or relaxed, in the future if one is faced with a different set of circumstances. Obviously the assumptions should not be so extensive so as to make a theory overburdened with constraints and subsequently meaningless.

Once this stage has been completed, one can construct an economic *model* of the issue under investigation such as the housing market. The resultant theoretical model should, in this example, be able to explain the workings of the housing market, but it should also enable us to predict the future of that market if it is exposed to such stimuli as a change in interest rates, or a change in mortgage related tax relief, and so on.

To ensure that our model is operational and is able to explain present circumstances, we should continuously subject it to *tests* using either *primary data* (data collected at first hand, such as data obtained from questionnaires or observation), or *secondary data* (data collected from other sources, such as property journals, or government statistics).

After extensive testing the *results* should be set out, with the aid of descriptive statistics, in a clear manageable form. Having pages of un-summarized, un-collated data will be of little use. Again the decision may have to be made concerning which data to include and which to leave out. At this stage care should be taken to ensure that although descriptive statistics may promote an understanding of the results, they can be, and often are, misused so as to place a certain emphasis on, or change the true emphasis of, results.

Once the results have been presented and assimilated, a careful process of *analysis* needs to be embarked upon so that the model can be used in a meaningful manner, and can for example be used to forecast the future of the market given a likely set of impending circumstances.

Finally the findings need to be *presented* in a clear, understandable, and *summarized* format that will encourage an end user to actually utilize the model , or theory, for their benefit.

The important point to realize here is that theory and reality are *not* two separate issues. Simply, good theory should enable one to understand the current and past workings of a market and to be able to forecast changes in that market in the future. Those who perceive theory and reality to be divorced could be suffering from any of the following scenarios:

1 They have failed to appreciate that theories are normally simplified so that they can be adapted for a wide range of uses, and are easy to use.

2 They have failed to realize that many theories are in fact designed to show extremes, rather than actual cases, so that we can compare real-life issues against these extremes.

3 They have been let down by a theory in the past and have thus become distrustful of theory in general. Such a reaction is understandable as theories can 'break down' and become less applicable if they are not changed with time as underlying circumstances change. However, it is often the case that the theory has been wrongly used or simply misunderstood in the first instance.

4 They may be operating in practice in ways not suggested by existing theory. This could mean that there is an alternative (better) theory available, or, as has been the case, firms, although surviving, are simply run incorrectly and are not maximizing their potential due to an ignorance of known theory and improved practice.

5 Firms may be operating along the same lines as suggested by the theory but are simply unaware that this is the case. This is unsurprising as theory, in part, would be governed by how these firms operated in the first place. For example, a theory about house prices may also take on board the behavioural attitudes of estate agents, and house builders for example.

However, the following *potentially* negative points of theory must also be clearly understood:

1 One must remember right at the outset that, as with all subjects, one should view what we learn in a *critical* way so that the process of theory and application can be improved over time. Specifically, if we never questioned existing ideas or practice we would never have evolved out of the 'stone age' for example! Moreover, if we took another science, say physics, as our example, we can see how major new ideas have changed the way that we view the world most dramatically, and even now it would be foolish to assume that we have fully understood the workings of the universe. If we did, we would be able to forecast the weather hundreds of years in advance. Furthermore, there are some who advocate *chaos theory*, which essentially says that nothing can be predetermined or forecast due to unexpected and even unaccountable change.

To emphasize this point yet further, we should be very careful not to cling on too eagerly to existing theories if they no longer work very well as models to help us explain and forecast. For example: a model designed to describe the housing market one hundred years ago is unlikely to take into account modern variables that would need to be included in a model for the market of today. Faced with changing circumstances one must decide whether we should merely adapt existing theories to take into account any new facts, or rethink the whole concept again. At this juncture, one should bear in mind that continually adapting a theory could 'overload' it and make it unnecessarily cumbersome and eventually unworkable.

2 It must be appreciated that the success of any science rests upon the initial ability of scientists to separate their observations on *what does happen*

from their views on *what they would like to happen*. For a dramatic example one only has to be aware of the uproar that Charles Darwin caused when he introduced the concept of evolution as this did not fit into the biblically orientated interpretation of the world that scientists had forced their theories to fit in the past.

One way of reducing this problem is to distinguish between **positive** and **normative** statements as we should be careful not to make normative statements without first undertaking positive science. Essentially a positive statement is one that can be **tested** using past and present data, whereas normative statements are concerned with what 'ought to be', such as judgements of good and evil. As such, the latter type of statement is invariably bound up with cultural, philosophical, and even religious viewpoints, which all depend upon individual value judgements.

An example of a positive (testable) hypothesis would be:

'The higher interest rates are, the lower house prices will be'

In this instance data could be obtained to see if there was a link between the two variables. Whereas, a normative statement would be:

'The Chancellor should not hurt homeowners by using the interest rate as a macroeconomic weapon'

Such a view will obviously depend upon your own value judgements.

However, even with positive statements we must be careful that we do not misuse the data before us so that we disprove or prove our hypothesis depending upon our personal views. You should never try to prove your hypothesis, rather you should *test* it to see whether it is right or wrong. Moreover, this is important even at the point of data selection, that is one must avoid the temptation to selectively choose data to prove a point. Furthermore, one should not be frightened to test issues merely because they do not conform to the views of current practice.

3 Just as with other sciences, it would be ideal to gain observations via controlled experiments, but obviously this cannot happen when we are looking at something as complex as 'society'. For example, if interest rates were to change, we cannot hold all other variables constant to see their impact upon consumer spending, as it is highly likely that many other variables will be changing at the same time. For example: even the time of year has an impact upon consumption via seasonal spending. However, in economics, in an attempt to isolate the influence of other variables we *temporarily* assume that all other variables are held constant. When we do this the term ***ceteris paribus*** is often used, or assumed, which simply states that all other variables have been held constant. In the following sections of this book this point is assumed unless otherwise stated.

4 Finally, we must tackle the criticism that economics will not work because of the individuality of human behaviour. Just by looking around us, evidence certainly does show us that there are individuals in society,

but, on average, human behaviour is relatively predictable. For examples: hot weather increases orders for outside patios and barbecues to be built, and more people make use of leisure property such as swimming pools and golf courses. Furthermore, obviously norms do exist in human behaviour as otherwise complete chaos would reign as life would depend upon people's whims from day to day. Therefore, we can normally predict things with a ninety-five or ninety-nine per cent *level of confidence* depending upon the specific matter under examination. In fact many assume the existence of the 'rational economic man'. In other words, as the majority of people behave in a certain manner, one can take this general behaviour as the 'norm'.

To finally drive these issues home, a key point to appreciate is that although economics is still in its relative infancy compared with more established sciences such as physics, it still has a solid foundation of thought stretching back for hundreds of years. For example, one just needs to mention a few of the great names in economics to see this development over time:

John Locke	1632–1704
Adam Smith	1723–1790
Thomas Malthus	1766–1834
David Ricardo	1772–1823
John Stuart-Mill	1806–1873
Karl Marx	1818–1883
Alfred Marshall	1842–1924
John Maynard Keynes	1883–1946

Not to forget the great philosophers such as Popper, Lakatos, and Kuhn who were the true source of much of our subject theory.

PART ONE

'Market analysis' and its application in the built environment

Obviously in an economy where *all* decisions are made by a central governing body, the number of buildings built in any given year, their location, the type of building, the price at which they are sold, or rented, and so on, could be planned in advance, and, subject to no unanticipated constraints, results should be achieved. The relative merits and demerits of such a **centrally planned** system require separate and careful debate, but it is quite obvious that in most countries such a system has not been, or is no longer, adopted in the main. Rather, it is felt that such decisions as those listed above, for example, can, in most cases, be automatically and efficiently dealt with via an interaction of consumers' and firms' demands made upon suppliers. That is, we let the **market system** dominate. However, it must be noted right at the outset that, as we will be seeing later in more detail, it is sometimes felt that the market does not always give us the best solution. Specifically, problems can arise because of a situation known as **market failure**, whereby it is felt that we can attempt to correct market failures with local or national economic policy, (see Chapter 15). For example, this can be exhibited by town planners ensuring that there are sufficient 'green' areas in and around our towns and cities. Such intervention in the market gives us a **mixed economy**, where the mixed economy allows the market to dominate but will attempt to correct its failures via public intervention.

However, despite market failure, it is rarely total, and thus the market is a powerful and dominant force. Therefore, a thorough understanding of how markets work will enable one to apply the concept to explain, for example, why certain buildings are built, why a particular construction process was used (subject to technological constraints), what will happen to the level of commercial rents in a particular area, or house prices in the future, and so on. That is, the analysis is prescriptive as well as being descriptive. Indeed one can even go so far as to assert that the architecture of the day is strongly influenced by market forces.

Thus, what follows is a thorough introduction to the mechanism of the market aimed at the individual who wishes to apply the principles, and is not fundamentally concerned with the deeper theoretical underpinnings of the concepts used. Despite this, though, this section is a thorough explanation of what is required for practical application, accompanied by a good degree of academic rigour.

Initially, the section will confine the discussion to five different markets, but at the same time it must be emphasized that such analysis is remarkably versatile and could be used to explain all manners of goods and services in the economy. This point will hopefully be appreciated as you progress through the book as a wide range of applications in the built environment are examined. Because of such a broadness of applicable scope it is difficult to select just a few examples, but the ones that have been chosen will serve to show several important aspects of different categories of market. After understanding the content of this section it would be a useful exercise for you to see if you can apply market analysis to explain other real estate related markets of your choice.

In order to easily familiarize the reader with the concepts, and to demonstrate a common day practical usage, we will examine the rental market for student accommodation at university. This is likely to be a market that many readers of this text are directly involved in as tenants. After this 'foundation laying' exercise, the book goes on to develop the analysis further so as to apply the theoretical concepts to a range of construction case studies. Namely:

1 the owner occupied housing market;
2 the market for building land;
3 the market for construction components;
4 the market for building labour.

Finally, Part 1 will briefly examine the rationale and subsequent implications of government activity in construction and property related markets.

1

Introductory market theory

Probably the best way to understand how a **market** works is to initially split it into its constituent parts of **demand** and **supply**. This approach will enable one to examine the forces that determine and affect these two key elements as they work together in determining both the price of goods and services and how much of them are traded in the market. Then, as this section progresses, we will go on to introduce more complicated issues such as the concept of **elasticity**, before going on to consider some applications of the theory to issues related to the built environment.

(A) Demand

Right at the outset we need to assert that when we refer to 'demand', we are specifically implying what is known as **effective demand**. That is, economics is interested in one's demand for a good or service *coupled* with the ability to pay for that good or service. If we did not limit our studies to effective demand we would waste valuable time discussing people's desires to own unattainable assets. For example we may all demand to live in a large house and have a luxury car, but these demands are irrelevant if we cannot influence the market by actually purchasing these items in the first place. Therefore, effective demand is assumed, although normally just the term 'demand' is used by itself in economics so as to save writing or saying 'effective demand' each time we discuss or utilize the principle.

Although there are perhaps some goods that we buy unexpectedly on impulse, before we purchase most goods and services we normally consider a variety of variables that would help us in the decision process of whether to buy or not to buy. However, because we make purchases so frequently (try to count how many goods and services that you have bought in the last ten days for example) we may not be consciously aware of the thought processes that are occurring in our mind. Normally, though, such thought processes do become more obvious if we are making a purchase involving large sums of money such as is the case when buying a house, or household contents such as furniture and carpets for example. Important variables that we may consider during the purchasing process are now given in the following list:

1 the price of the good in question (P);
2 one's income (Y);

3 the price of other related goods and services (S);
4 one's tastes or preferences (T).

As these variables are referred to frequently a simplified notation form is given in brackets after each. This section will examine each of these influences upon demand in detail, but it will become apparent as our analysis develops that other factors are also of great importance in determining the level of demand. The list of additional factors is nearly endless, but perhaps the main ones to consider are:

1 demographic factors (D);
2 expectations of future events (E);
3 government policy (G).

If we were now to express the relationship between demand and the above mentioned factors in 'functional form' we could write in the most simple of cases:

$$Qd = f(P,Y,S,T)$$

In other words, the quantity demanded of a good or service is normally a function of: price, incomes, the price of other goods, and one's personal tastes. The way that the demand function has been written, and the letters chosen as abbreviations of the variables, forms the word *pyst*. This word has been 'designed' so that if you can remember it you should be able to recall, and therefore analyse, the key determinants of demand for any good or service that you need to investigate whether it be the demand for 'starter homes', the demand for building land, or the demand for bricklayers for example. This gives us the standard demand relationship, but if we were now to expand this function to consider the other variables mentioned above we could write:

$$Qd = f(P,Y,S,T,D,E,G)$$

Again it must be stressed that this is by no means an exhaustive list of all the variables that may have an impact upon demand. For example it may be the case that the demand for some goods and services is affected by such considerations as climatic conditions and even 'brand loyalty'. Despite this, the above functions do take the major variables into account in most instances, and thus they enable us to have a comprehensive look at most issues. In addition to this it is quite normal to add an 'error term' (u) at the end of the function to represent the fact that not all of our demand observations will be totally explained by our suggested variables. Moreover, one can *weight* each variable by means of a coefficient in order to show its relative importance as an explanatory variable.

A case study – the demand for rented student accommodation at university

This case study examining the demand for rented student accommodation at university has been selected as a useful illustration of the theory of demand for two reasons:

1 The majority of people using this book will be attending a university or a place of higher education, and will probably have already experienced, or will be experiencing, at 'first hand' this particular market at work as they find suitable accommodation for the academic year.

2 As it is dealing with a sub-section of the property market it is felt that it will be a useful example to demonstrate how theory can be applied to a real-life issue in the built environment.

Although this part of the section is not meant to be a 'guide' about university accommodation, it will firstly set the scene of our 'case study' by looking at the possible different types of accommodation that are available to the student, and some criteria which the student may use when determining where to live. Each year hundreds of thousands of people will enrol to undertake courses at universities throughout the world. The majority of these people will be young school leavers, but there will also be some older people joining most universities as mature students. Therefore one of the first points to realize is that markets have to cater for different client groups, and that each client group may consider a different range of variables in their demand function, or at least place different values on each variable. For example, the young school leaver may wish to live close to areas of entertainment, whereas the older student may value peace and seclusion when deciding on where to live. However, whatever their age, sex, or nationality, the one thing that they will have in common is that they will require some form of acceptable living accommodation whilst they are studying at the university. The choice of accommodation that a student can consider is normally quite wide:

1 Most towns that contain a university will have a substantial stock of private rented housing that is aimed at student occupation. Landlords will appreciate that by dividing their properties up into units available for multiple occupancy, they can derive a satisfactory return from their investment in lettable housing. Typically, in such situations, four or five students would share a large house together.

2 Some towns may have properties owned by 'Housing Associations', or similar, that specifically cater for people with low incomes. As students often fall into this category they may be successful in renting such a property.

3 The majority of universities can also accommodate a large number of students in their own purpose-built 'halls of residence'. Such buildings normally consist of a large number of study bedrooms with communal facilities such as kitchens, and entertainment areas. Some halls of residence are of the 'traditional format' whereby meals are served, whereas others are exclusively self-catering.

4 Some students may live close enough to their university to be able to live at home, or they may have relatives or friends in the vicinity with whom they can stay during the term time.

However, it is not only the type of dwelling that is an important consideration. Other factors have to be taken into account when the student assesses the suitability of the accommodation. Factors that will be looked into, and the degree of importance placed upon them, will depend upon the individual in question. Possible considerations are given below although this is again by no means an exhaustive list:

1 Is the accommodation close enough to the university to make it an easy or pleasant trip into and out of work?

2 Alternatively, is the accommodation sufficiently far away from the university so that one can enjoy a break from the working environment?

3 Is the area safe at night?

4 Do any friends, or other people on the same course, live in the area?

5 Are there a sufficient range of facilities and social amenities in the area?, and if not are they easy to get to? Facilities that one may wish to have near one's place of residence would be: a laundry, cinemas, shops, bars, sporting provision, railway station, and restaurants to name just a few.

6 Is the area sufficiently serviced by public transport thus making journeys to university, town, or home readily accessible?

7 Is the quality of the accommodation of a satisfactory standard? Many areas may have an abundance of cheap rooms to let but they may be in an unsatisfactory state of repair.

8 Are there any other costs of occupation? For example, are the bills for services such as electricity, gas, and water, paid for by the landlord or the tenant?

Now that we have an overall picture of the market in question we will now narrow our study down to looking at *one* of the types of student accommodation listed above namely: *university owned halls of residence*. This is effectively a **sub-market** of the overall market for student accommodation and has been selected as it is one of the most preferred choices for new students. It is assumed that the university operates its halls of residence on a pure market principle and wishes to obtain profits from their operation, rather than artificially interfering in the pricing and allocation of the rooms, on grounds of equity for example. In order to appreciate the various forces at work that determine the level of demand in this market we will now examine, in initial isolation, each variable of our demand function given above. It is felt by many that the most effective and understandable method of presenting such information is via an explanatory (demand) diagram. Not only do diagrams help to summarize key issues , but they are often a welcome break from potentially tedious text in essays and reports. In drawing your own diagrams please note the importance of giving the graph a title, and ensure that you fully annotate the axis and any curves that you have drawn. This ensures that people can immediately identify the issues that you wish to portray.

Before we examine these issues with respect to our first example, I must again briefly remind you of the importance of the term *ceteris paribus*, which simply means that in order to attempt to isolate the relative impact of any of these variables we need to temporarily assume that all other variables are held constant. Thus, if we are looking at changes in price, for example, *ceteris paribus*, it means that we *temporarily* assume that all the other variables in the equation, such as incomes and tastes, have remained the same. This is analogous to a chemistry experiment that requires a *control*. Obviously in chemistry it would not be sensible to do something different every time you tried to repeat an experiment as the results are likely to be different due to other influencing factors. Likewise, in economics, if we are considering, for example, a change in demand for detached housing due to a rise in personal disposable incomes, we temporarily hold all other variables constant so as to isolate the pure effect of the income changes. One could then go on to investigate other variable changes if and when they occurred. Therefore, throughout the following analysis of individual variables it is assumed that this condition of *ceteris paribus* holds although it is not specifically stated.

(1) The demand for halls of residence and the impact of price (*P*)

For most of us who do not have access to extreme wealth, price is perhaps the most dominant variable in the decision-making process. Simply, if 'hall fees' (rents) are relatively cheap, in comparison to other forms of accommodation, halls are likely to attract a high demand as students try to secure a place in such accommodation. Conversely, if hall fees are too high it will discourage people wishing to live in a hall and they will attempt to seek alternative accommodation elsewhere. Therefore, at low prices we would expect a high demand, whereas at high prices we would expect low demand. As such, an *inverse*, or *negative*, relationship exists between the price of a good or service and the quantity demanded of that good or service. This can be seen diagrammatically in Fig. 1.1. Here it can be seen that because of the inverse relationship between price and quantity the demand curve is downward sloping when viewed from left to right. This implies that with the existing level of demand, if our hypothetical university sets rents as high as R_3 they will only be able to rent out 400 hall places. At the other extreme, if they only charge a rent of R_1 they may attract 1000 applicants for hall places. Those of you who have a good mathematical training will realize that essentially the graph is the wrong way around. Standard mathematical rules of consistency tell us that the **dependent variable** should be on the y, or vertical axis, and the **independent variable** on the x or horizontal, axis. This is not the case here as we have seen that quantity demanded is dependent upon price, and as such is the dependent variable, yet it appears on the x axis rather than the y axis. The simple reason why it is the wrong way around is that many of the early market economists did not know much about the rigours of mathematics. Therefore, it is because of this early 'mishap' that tradition has left us with a 'backwards' diagram,

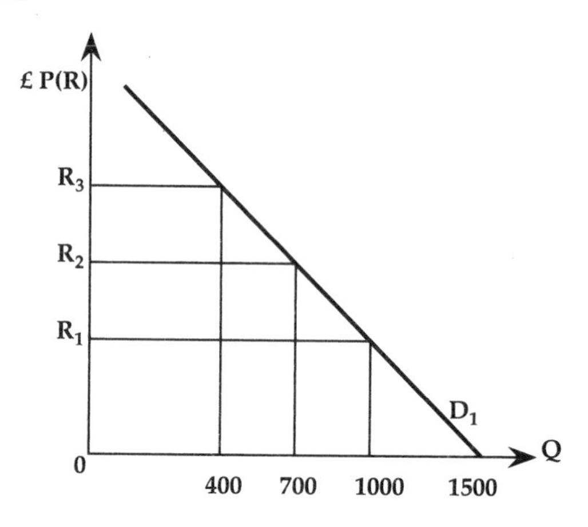

Figure 1.1. *The demand for halls of residence*

and it is in this format, as shown in Fig. 1.1, that demand is always represented. You should appreciate though that this non conformity with mathematical rules makes no difference to the accuracy of the analysis.

Soon (see Chapter 2) we will be introducing the concept of elasticity to our analysis so that more specific statements and predictions about demand can be made. However, at the moment, it is sufficient to say that there will obviously be some limits on such demand. For example, even if the university offered free hall places there will only be a finite number of students attending the university, and many of them may still elect to live elsewhere as they prefer the attributes of alternative accommodation even at a price. Such people may be willing to pay a rent in the private sector if such private sector accommodation is nearer to sports and social facilities for example. Our diagram suggests that only 1500 students would take up the offer of free hall places. Despite this it is true to say that the quantity demanded of a good or service is largely dependent upon the price of that good or service. As we have observed different numbers of students expressing an interest in hall places as price changes by simply reading off the figures from the demand curve, the essential conclusion on the topic of price is that: *any change in the actual price of a good or service will cause a movement along the existing demand curve, and will not cause the demand curve to move.*

However, pure price changes should be distinguished from **expectations of price changes** as these can cause the demand curve to actually move (shift). For example, imagine that it was rumoured that hall fees were to rise substantially in the next academic year. This fear of rising prices may encourage many to start looking elsewhere for alternative accommodation. If these people are successful in their quest for finding somewhere else to live they will no longer require a hall place in the new academic session. Therefore, their demand has been withdrawn from the hall of residence market and has gone to another market. Thus, *fears* of rising rents would have the effect of shifting the demand curve to the left as fewer places in halls of residence are demanded at each potential price (rent). Similarly, a fear of future rises in rents in the private rented sector could create an increase in current demand for the halls of residence as students rush to obtain a place in accommodation that is likely to be cheaper than the alternative in the future even if it is not so now. This behaviour is likely to cause a shift to the right of the demand function for hall of residence places showing that more rooms will be demanded at any given price as students leave the private rented sector in preference for the university owned accommodation. Fig. 1.2 shows such an increase in the level of demand as the demand curve shifts from D_1 to D_2. Notice that whereas previously 400 hall places were demanded at a rent of R_3, 600 more people have now transferred out of other forms of accommodation and are willing to rent a room in a hall of residence at this price. Thus, the total level of demand at R_3 has increased from 400 to 1000 people merely because of expected price changes, which may or may not occur, rather than actual price changes.

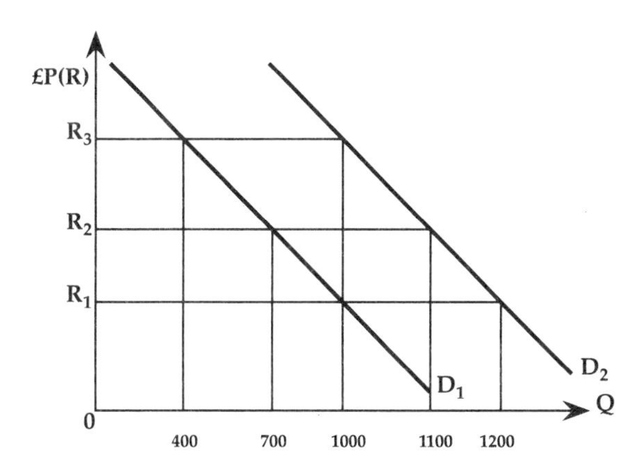

Figure 1.2. *Demand shifting to the right'*

(2) The demand for halls of residence and the impact of income (*Y*)

Before we return to the case study of university halls of residence a few general issues regarding the level of demand and income will be discussed. As has been mentioned earlier in the section, economics is only concerned with effective demand. That is, demands must be backed up with the consumers' ability to pay for the goods or services that they require. Thus, it is obvious that **income** is a crucial deciding factor in determining how many goods and services will actually be purchased. More specifically we should be concerned with the level of **disposable income**. That is, how much money consumers have at the end of the day net of all deductions from their earnings such as income tax, national insurance contributions, payments to superannuation pension funds, and so on. One can express such disposable income in terms of a simple formula as shown by:

$$Yd = Y - (Ty + Dy)$$

where: Yd = Disposable income;
Y = Income from all sources;
Ty = Income tax dependent upon your level of income;
Dy = Other deductions taken at source, such as national insurance, that are also dependent upon income (y).

Re-writing information like this via the use of a simple mathematical expression is certainly not essential, but it is considered by many, as with the earlier example of demand being a function of *pyst*, to be a useful summary tool, as well as an easy way of remembering things as one only has to learn one expression rather than several paragraphs of text saying the same thing. Moreover, you can extend such terminology to help understand how much 'spending power' consumers have 'left in their pockets' with which to order goods and services after one has also taken into account other deductions that are likely to occur from a consumer's income. Other payments which have to be made from one's earnings could include items such as monthly mortgage repayments (*m*), payments into endowment policies (*ep*), home and contents insurance (*hci*), car insurance (*ci*), and repayments on bank or credit card loans (*b*) for examples. Taking all of these payments into account would enable one to calculate how much income consumers have left over (*Yp*) for other expenditures such as those on food, drink, and entertainment. This calculation is often performed if you need to apply for a mortgage or a bank loan as it should give you an indication as to whether or not you can take on any further financial responsibilities. Adding these factors to our original equation above will give us:

$$Yp = Y - (Ty + D + m + ep + hci + ci + b)$$

where Yp = money left over in one's pocket.

When increases in disposable income occur, say due to a reduction in tax, or an increase in pay levels, consumers are likely to buy more of what they consider to be **normal goods** as they can now afford to increase their consumption. Therefore, under such circumstances of rising incomes, the demand for the majority of goods and services would increase, and therefore the demand curves for these items would shift to the right. Conversely, it should be evident that a decline in purchasing power, say due to an increase in the rate of interest filtering through to larger mortgage repayments for homeowners, consumers will be forced to buy less at any given price. Under such conditions of declining disposable incomes the demand for normal goods would drop and thereby shift to the left. However. it should be noted that not all goods and services can be categorized as being 'normal goods' as some may be perceived by consumers as being **inferior**. Inferior goods and services are those that consumers will tend to view as only being purchased out of necessity, or because they are cheap when one's income is low. For example, 'bargain priced assemble-it-yourself' furniture, and even rented accommodation may both be viewed as inferior goods. If this were the case one would expect the demand for these goods to decrease (shift to the left) as incomes increase rather than the other way round as would be the case if the goods were viewed as normal. Conversely, the demand for such inferior goods is likely to rise (shift to the right) as people's incomes drop. This seemingly 'perverse' behaviour can be explained by the fact that in periods of falling incomes fewer people can afford the luxury of owning their own house or purchasing expensive home fittings.

It should also be mentioned that **expectations** are again an important influence upon demand. The reason for the linkage between expectations and income is that according to **Friedman's Permanent Income Hypothesis**, many will purchase goods and services depending upon their **expected future income** rather than merely basing their consumption decisions upon present incomes. Thus, if one was expecting a forthcoming pay rise, for example, the easy availability of credit, or the running down of previously accumulated savings, would enable one to consume more *now* simply based upon the knowledge that such consumption behaviour can be paid for in the future, or that one's savings position could be reestablished at a later date. Thus, assuming normal goods and services, such optimistic expectations of increasing incomes can cause the demand curve to shift to the right as if incomes had already risen. Conversely, if one is pessimistically anticipating redundancy, or is expecting a significant downturn in future disposable income due to an on-going recession, for example, one may decide to 'save for a rainy day' and cut current consumption now, although present income is, as of yet, unchanged. Therefore, in such circumstances, the individual's demand curve would shift to the left. The amount by which the curve would shift to the left would depend upon whether the consumer viewed the good in question as a **luxury** or as a **necessity** as it is likely that the consumption of luxuries can be reduced, whereas there is little one can do about the purchase of essential goods and services (see Chapter 2).

Applying this to our case study of a university hall of residence we could imagine the following possible changes in demand for such accommodation if student incomes were to change. Increasing student incomes could occur if government were to give students more financial assistance for undertaking courses at university, or alternatively, a student's spending power could also be enhanced if more loan finance were made available to them. If this were to occur we may expect more students wishing to live in their own rented accommodation rather than living with parents or relatives, or fewer students having to share with friends. This is likely to have the effect of increasing the demand for most types of rented accommodation, and thereby shifting the demand curve to the right. An increase in demand for places in a hall of residence would depend upon whether that accommodation is perceived as being 'normal' or 'inferior' in nature. If other types of accommodation, such as houses in the private rented sector, for example, are viewed as being inferior, say due to their comparatively poor state of repair, or location, an increase in student incomes is likely to encourage more students to apply for a hall place and thus the demand curve for 'halls' would increase by shifting to the right. On the other hand, if student incomes were to fall, we would expect an increase in the number of students having to live at home, or to share a house and live in cramped conditions. Furthermore, many would have to accept accommodation of a poorer, cheaper standard. This would lead to a drop in demand for most types of rented accommodation (a shift to the left of the demand curve) as there would be an increasing tendency for multiple occupancy. However, there may be an increase in demand (shift to the right) for some inferior properties at the 'lower end' of the market.

(3) The demand for halls of residence and the price of other goods and services (*S*)

Before returning to the case of student accommodation some general points need to be appreciated: essentially, in determining the demand for any good or service, say owner occupied housing, one must also take into consideration the price of other goods or services on the market. Firstly, it is highly likely that there are alternatives such as rented flats, or public sector dwellings, for example, that one could choose from as a means of acceptable shelter. In other words there are **substitutes** (hence the letter *S* in our demand equation) to the original good, or service, in question which the consumer may view as having similar attributes. Therefore, the price of such substitutes may well influence the consumer's purchasing decision. Secondly, it is likely that there will also exist **complementary** goods and services which are items that are normally consumed in conjunction with the original good or service. The price of these complementary items may have a bearing on the consumer's ability to buy a particular good or service as they represent additional costs that have to be incurred on purchasing the original item. For example, by purchasing a house one may be liable for a local property tax payable by people who own property in a particular area. Such a tax has to be consumed in conjunction with the house and thus represents an additional cost of occupation. Moreover, the price of other goods in general is an important issue as they could effect a consumer's overall disposable income. For example, if the price of goods increases one will generally have less money left over to effectively demand any good or service that is desired. In order to assess the importance of both 'substitutes' and 'complements' the book will now briefly examine each in turn in relation to the case of student accommodation

(a) The price of substitutes

Many students may perceive that a room at a university hall of residence is a very close substitute to a room in a rented house in the same area. However, even if the two forms of accommodation were available at the same rent it is highly unlikely that they would be seen as *perfect* substitutes. The reason for this lack of perfect substitution is that no two buildings have identical attributes as they will differ in a variety of ways ranging from their exact location, and state of repair, to their 'general atmosphere'. Despite this lack of perfect substitution there are still sufficient similarities between the two so that a change in the price of one may well effect the demand for the other. In other words the relative prices between the two goods is an important consideration for the consumer in deciding which one to purchase, or rent as would be the case in this example. For example, if the university were to increase the rents for their halls of residence this may well lead to a decline in demand for hall places and an increase in demand for rooms in rented houses. In terms of a simple mathematical relationship we could express this as:

$$\uparrow P_H \rightarrow \downarrow Q_{DH} \rightarrow Q_{DR} \uparrow$$

where P = Price;
Q_D = Quantity demanded;
H = Rooms in halls of residence;
R = Rooms in a rented house.

That is, there is a *direct* and *positive* relationship between the price of the original good and the quantity demanded of the substitutes. For example, as we have seen, an *increase* in the price of the original good (halls of residence) has led to an *increase* in the demand for the substitute (rented houses). The strength of this relationship between the two goods can be measured by calculating their **cross price elasticity** (see Chapter 2). Moreover, it will soon become clear, after we have examined the supply side of the market, that such increases in the demand for the substitute good could increase its price, and that such price increases could 'put the break' on such substitution as the relative price differentials between the two goods is eliminated. Diagrammatically this relationship can be seen in Fig. 1.3. In Fig. 1.3 it can be seen that as the rents in halls of residence are raised from P_1 to P_2 the number of people looking for a place in hall falls from Q_1 to Q_2. As these people (Q_1,Q_2) still need to find a place to live this will lead to an increase in the demand for alternative forms of accommodation such as that found in the non-university rented sector. Therefore, one would observe a shift to the right of the demand curve for non-university sector rented accommodation as seen by the movement of the demand curve from D_1 to D_2. Similarly, a *decrease* in rents in the non-university sector would increase the demand for such accommodation yet cause a *decrease* in the demand for halls of residence as they became relatively more expensive. Therefore, yet again, in the case of substitutes, there is a direct and *positive* relationship between the price of one good and the quantity demanded of another as both move in the same direction. In terms of a simple mathematical expression, and using the same abbreviations as above, this can be seen as:

$$\downarrow P_R \rightarrow \uparrow Q_{DR} \rightarrow Q_{DH} \downarrow$$

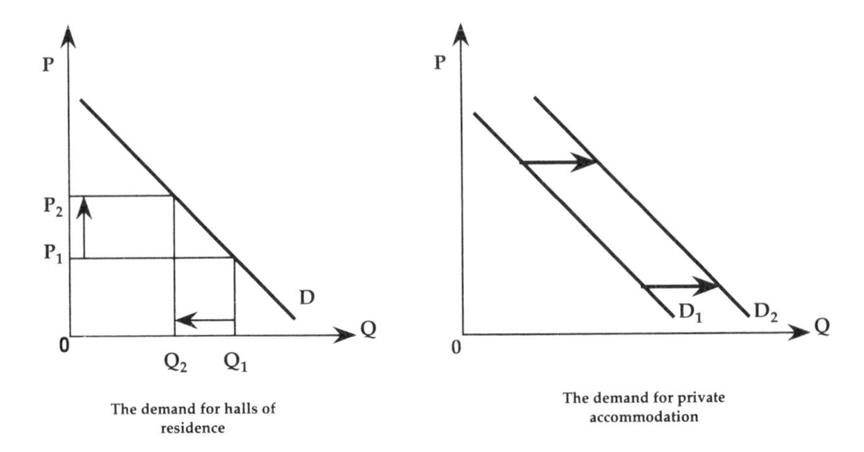

Figure 1.3. *The impact on a substitute good due to a change in price of the original good*

(b) The price of complementary goods and services

Complementary goods and services are simply those goods and services which one tends to 'consume' in conjunction with the original good. Continuing with the example of student accommodation, other additional items that a student may have to pay for in conjunction with renting the room are: local property taxes, and bills for services used such as gas, water, and electricity. Therefore the relative cost of these items should be taken into consideration when purchasing, or renting as is the case here, the original good. For example, a modern, purpose-built hall of residence is likely to be more energy efficient than an old converted house with draughty windows and a high ceiling. With such a comparison any additional bills for heating, for example, are likely to be significantly lower in the hall of residence as opposed to the old house in the non-university rented sector, making the former a more attractive overall financial package. If we now view the non-university rented accommodation as the original good under consideration one can see that a rise in energy prices would encourage people to cut down on their consumption of energy wherever possible, but moreover, they would be inclined to vacate energy inefficient dwellings in search of more modern, better insulated accommodation. Therefore, in the case of complementary goods there is an inverse relationship between the price of the complement and the quantity of the original good. In other words as the price of energy increased, the demand for old rented accommodation of poor quality would decrease as the demand curve for such properties shifted to the left. Alternatively, this can again be viewed as a simple mathematical expression:

$$\uparrow P_E \rightarrow \downarrow Q_{DE} \rightarrow \downarrow Q_{DR}$$

where E = energy.

Notice though that this relationship is not always that clear in terms of a property market such as the one under consideration as we would have seen a positive relationship between the price of energy and the demand for the energy efficient halls. Here it would be an extreme case whereby energy prices were so high that people would be forced to try and reduce the overall cost of occupation by sharing a room, or living with friends or relatives for examples. Once we have reached this extreme we would witness the expected inverse relationship:

$$P_E \uparrow \rightarrow \downarrow Q_{DE} \rightarrow Q_{DH} \downarrow$$

More straightforward examples are usually put forward in text books with goods that have no close substitutes that could complicate the analysis. For example a common example of complementary goods is that of petrol and cars because in most instances one good is of little use without the other. In such a case one would expect that with very large increases in the price of petrol the demand for cars, and therefore petrol, would eventually decline. In other words we have the expected inverse relationship between the price of a complementary good and the demand for the original good. Even in this simplistic example though, research has revealed that the reaction to high energy prices varies over time. For example, increases in petrol prices have encouraged people to be initially more economical with the way that they drive and the distances that they travel. Indeed sustained high energy prices have encouraged people to look for smaller, more economical vehicles. However, research also shows that, in the long run, people become used to the new relative set of prices and begin to consume in the same way as before. Thus, even with high energy prices we still have the phenomenon of 'high performance cars', and the mass use of the personal car for example. In conclusion, therefore, one has to be aware that 'complementarity' induces forces that would lead to an inverse relationship between the price and the quantity of the two goods, but the existence of substitutes exerts forces in the other direction. The overall effect will thus depend on the relative strength of each force.

(4) The demand for halls of residence and 'tastes and preferences' (T)

Tastes and preferences represents the last of our four basic demand variables, although by no means the least significant one. However, although it is an important category 'tastes and preferences' are difficult to measure as they can be influenced by such a wide variety of factors ranging from: advertising, brand loyalty, and peer group pressure. Essentially if something is fashionable we would expect an increase in the demand for such a good or service which would lead to its demand curve shifting to the right as more of the product is demanded at each potential price. Conversely, if something is no longer in vogue, its demand will drop and thus the demand curve for it will shift to the left as less is demanded at any given price. Therefore, if, for example, students viewed halls of residence as an inferior form of accommodation for 'new' students, there may be social pressure to rent an alternative type of dwelling as it is the 'in thing' to do. Advertising is aimed at influencing tastes and preferences and much research has been done on its effectiveness. If, in our example, the university were worried that they may not be able to fill their halls of residence say due to cheaper accommodation being available elsewhere, they could positively advertise the halls of residence by promoting their proximity to the place of work, and negatively 'advertise' the poor attributes of alternative accommodation such as possible higher fuel bills.

(5) The demand for halls of residence and other influencing variables

For most goods and services their demand is also influenced by many other variables such as government policy and demographic change, for example. Therefore, in any study of demand one has to decide upon which variables are important enough to consider as significant determinants of overall demand. In the case of university halls of residence we could expect an increase in demand for such accommodation due to a variety of reasons such as:

1 Increasing financial support to students in the form of loans or grants. This may enable more people to meet the costs of higher education and encourage them to apply for a university place. Obviously any increase in student enrolment numbers will put additional pressure on the demand for accommodation. Such financial assistance may also help existing students afford better accommodation in terms of both quality and the number of people sharing. Therefore we are likely to see movements within the market as well as additions to it.

2 Local authorities may achieve the results described in (1) immediately above by giving students financial assistance in the form of a 'rent rebate' for example. A rent rebate is sometimes issued in circumstances where the council feels that the rent being charged is unjustly high for low income groups such as students. As such, the council contributes to some of the cost of the rent in the form of a refund.

3 Demographic changes, perhaps via a previously high birth rate, or migration into the area, could mean that there are more people of school leaving age who are seeking a place in higher education.

The individual demand curve and the total demand curve

So far we have examined the demand for accommodation in and around one university. Although this is useful for understanding the local market, it is often felt necessary to examine the whole market on a regional, national, or even international, scale so as to get an overall picture. In order to achieve this one needs to aggregate all the demands facing the various types of student accommodation for the geographical area in which you are interested. This could be done by seeing what sorts of accommodation are occupied by students at a range of given prices. In order to simplify this analysis for purposes of illustration, imagine that we were concerned with the demand for halls of residence at three universities in a region. The demands for places in halls of residence are shown in Fig. 1.4 for universities *A*, *B* and *C*. These individual demand curves are then aggregated to form the regional market demand curve, D_{MKT}, which illustrates the demands for such forms of accommodation in the whole region rather than just at one university. As one is simply examining how many rooms are demanded (along the x axis) in response to changes in price, or rent, (up the y axis) the process is termed **horizontal summation**. In our illustration it is seen that at university *A* 200 hall places were demanded at a high rent of P_1, yet 400 places were demanded at the lower rent of P_2. This extra demand as price is lowered is created by people who are expressing a desire to vacate their current accommodation and move into halls of residence as the latter has now become comparatively cheap. The demands facing other universities are likely to be similar, but not the same, as different conditions will effect different towns and universities. For example one possible explanation for the demand curve facing university C is that there is an ample supply of suitable alternative student accommodation in the town. As such, if the university were to raise its rents to P_1 students would simply find alternative, cheaper, places to live elsewhere. Shortly the concept of elasticity will be introduced so that one can discuss the differences found in these demand curves in more detail. However, it must be remembered that the problem with such analysis is that it may be quite difficult to find a significant price range in order to produce a demand curve, although this could be achieved by the universities 'experimenting' with rent levels for the purpose of the study. More importantly though, when dealing with property matters, it must again be emphasized that one will never be comparing an identical product. For example a hall of residence at one university may be identical to one at another university in terms of both design and structure, yet other attributes, such as location, will obviously be different. This is not the case with most goods as a particular vehicle, or type of computer, for example, are the same product wherever they are.

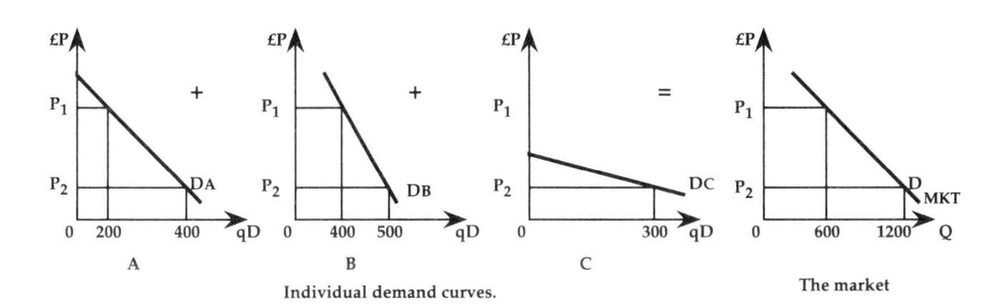

Figure 1.4. *Horizontal summation of individual demand curves in order to derive overall market demand*

(B) Supply

When discussing supply the key point to imprint firmly in your mind is that whereas demand comes from the consumer, or a firm demanding items from another firm, **supply** is concerned with the behaviour of the **producer** of a good or service, *or*, the **owner** of an asset, such as land, that is put up for sale. Importantly, it should also be realized that whether one is concerned with the renting of rooms to students, the sale of houses to owner occupiers, the sale of wooden window frames, or the sale of building land, as per some of the chosen examples in this text, or indeed any other good or service that one selects, again, as with demand, the essential principles of supply are the same for each market. Specifically, the amount of a good or service supplied will depend upon a variety of variables, the most important of which are listed below. As these variables will be referred to frequently a simplified notation form is also suggested in brackets when the term is first introduced.

Essentially, quantity supplied is normally dependent upon: price received (*P*); the aspirations, or expectations, of the supplier (*A*); the substitutes available to the supplier (*S*); the level of production technology (*T*); and input prices (*I*). You will note that I have used the abbreviation '*T*' to denote technology here, whereas it was used to signify tastes with demand. However, if you permit this inconsistency of abbreviated terminology it again gives you an easy function to remember:

$$Qs = f\,(P,A,S,T,I,)$$

In other words, if you can remember the word *pasti* it should enable you to recall and discuss each variable represented by each letter of the word. That is, quantity supplied is largely dependent upon the variables contained within the brackets of the function. However, as with the previous preliminary analysis of demand, although these variables are perhaps the main determinants of supply in most cases, this is not an exhaustive list. Other variables can be considered and their relative importance would depend upon the good or service, or asset, in question. Now, as with the analysis of demand, each of these variables will be examined briefly in order to determine their importance in influencing the level of supply.

(1) The influence of price (*P*)

The price that one receives for producing or selling a good or service is obviously an important factor that provides one of the greatest incentives to supply. That is, the more you get for your product, or assets, the more you are likely to supply on to the market as you will be receiving a greater reward for your effort or for what you own. Therefore, as price goes up, suppliers tend to want to supply more, giving us a positive relationship between quantity supplied and price. This relationship is exhibited in Fig. 1.5 where it can be seen that as prices rise from P_0 to P_1, for example, the

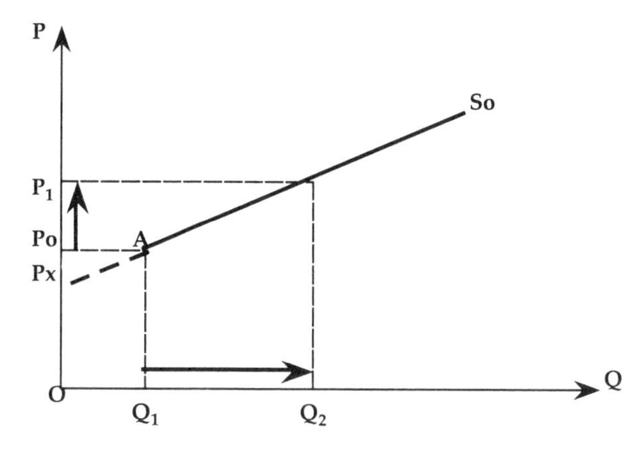

Figure 1.5. *A producer's supply curve*

amount suppliers are willing to supply on to the market increases from Q_1 to Q_2.

Returning to our example of the university owned hall of residence we can see that if the university could only charge a very low rent of *Px* it would not be willing to supply such accommodation as it is unlikely that it would be able to recover the administration, management, and repair and maintenance costs of such buildings. If however, demand was such that they were able to charge a rent of P_0, it would be in the institutions financial interests to provide such accommodation. Indeed if rents were as high as P_1 the university may be encouraged to build further halls as the profits received from such lettings could subsidize other key areas of the university's operations. Thus, there is more of an incentive to supply when prices are high as such high prices should ensure higher profitability and reward. One can observe this behaviour in all other areas of the built environment. For example, in times of high house prices major house builders attempt to acquire as many sites as possible in order to be able to sell as many houses as possible at peak prices. Conversely, when house prices are depressed in times of a recession, new building activity staggers to a virtual standstill in most areas. Therefore, it can be seen that as with the demand curve, *a change in price causes a movement along the supply curve*, whereas, as we will find out shortly, all other variables will actually cause the supply curve to shift. It is important to note that even though a manufacturer may cover costs at a price anything in excess of *Px*, they may not actually produce anything until a selling price of P_0 is achievable as the effort involved in production is not sufficiently rewarded at such low prices. Therefore, the supply curve would formally start from point 'A' as it is only from this point that production is of a sufficiently large scale (Q_1) to make starting it up worth while. This 'delayed start' of the supply curve for the producer potentially contrasts with the case of selling assets. For example: imagine the case of a large landowner in financial difficulty. If the person is a rural landowner they may have some land which is of poor agricultural quality which they are willing to sell of at any price in excess of *Px*, (or alternatively, the urban landowner could possess land with limited development potential due to either planning or site constraints). Yet other parts of the estate may contain prime agricultural land, and as such the landowner would need to be offered a higher price for it, say P_0 or P_1, in order to tempt them to part with it. (Likewise one could imagine the urban landowner with prime development sites in their asset portfolio.) In such cases, therefore, it is possible to envisage the supply curve starting from the price, or *y* axis. However, the curve would not come out of the origin as advertising and transaction costs such as estate agents fees and legal fees, would have to be covered from the sale receipts in order to make a positive return. Thus, a starting point such as *Px* would seem quite realistic. Furthermore, the supply curve would never originate from the quantity, or *x* axis, as this would imply that the landowner, for example, is always willing to give some of his land away for free. (If you are not sure about this final point draw a supply curve originating from the *x* axis and examine the implications of such a curve at different prices.)

(2) Aspirations, or expectations, of the producer (*A*)

Under this category we will examine the impact upon supply of the beliefs of those supplying the market concerning what will happen to the market for their product, or assets, in the future. For example, if the university sector were confident that student numbers, or financial assistance to students, were to rise in the future they may become *optimistic* about their ability to fully let a larger number of hall places. Such optimism about the future of 'the market' for student accommodation could encourage the university to either build more new halls of residence, or make more efficient use of existing space with their current building stock. Either of these options of expansion will increase the supply of rooms available in this sector. Therefore, hopes of an expanding market will often lead to an increase in supply which makes the supply curve shift to the *right*. Similarly, a firm manufacturing building materials may well increase its output in anticipation of a 'boom' in the construction industry as the firm would wish to have sufficient stocks in order to cater for the anticipated increases in demand for its products so that it too could take advantage of the construction industry's prosperity. Figure 1.6 shows that as supply shifts to the right it implies that more will be produced at any given price. One of the main reasons for such supply behaviour is that the firm will be confident that it can sell its stocks, but it may also wish to become more competitive so as to obtain a larger market share of the level of increased business. Conversely sales *pessimism* can cause the supply curve to shift back to the *left*. For example, if the university sector were to fear a decline in student numbers entering higher education they may cancel new building programmes, but may be able to reduce existing supply by converting some halls of residence, or at least parts of the building, to alternative uses. Likewise, if the building materials supplier feared an oncoming recession in the construction industry it may

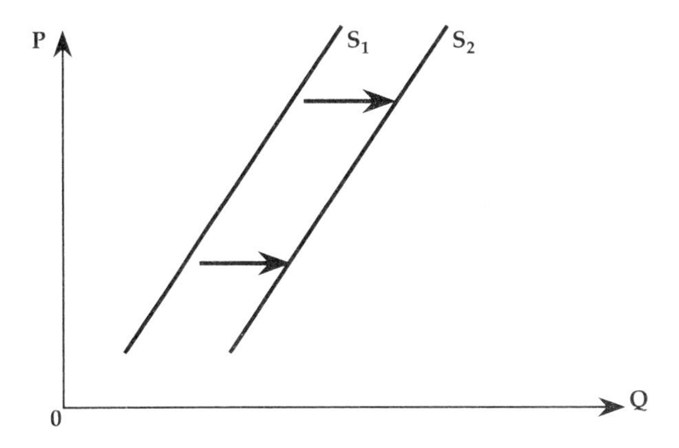

Figure 1.6. *Shifting supply*

well cut back output, in other words decrease supply, so as not to get left with unwanted stock that is expensive to store.

(3) Substitutes available to the producer (*S*)

In this context the term 'substitute' is used to imply any other product, or service, that the supplier could provide, and thus be involved in, if they were tempted to do so. This would obviously depend upon the entrepreneur's ability to become involved in other types of work. To understand the implications of such alternative forms of output it is perhaps best to use our example of the building materials supply firm. If the firm was currently manufacturing a particular type of roof tile, for example, it may find that there were more profits to be made by selling an alternative type of tile that had become more popular. If this were the case, and if it were technologically possible, the firm is likely to shift its output in favour of the new roof tiles. When this occurs the supply curve for the original roof tiles will shift back to the left as less are produced at any given price, whereas the supply curve of the new roof tiles will expand and shift to the right, as more of these tiles are manufactured at any given price. Similarly if the university found it more profitable to spend its buildings budget on conference facilities, less monies would be available for the upkeep and expansion of hall places.

(4) Technological change (*T*)

Technological change is also an important factor to consider. It should be noted that although most technological change can be viewed as being positive, in as much as it enables the producer to produce more efficiently, and thus causing a shift to the right of the supply curve, it can, in some circumstances, be negative. For example technological change can be reversed if improved health and safety regulations decide that a particular method of production is no longer considered to be safe. In this way, supply would be retarded to the left as a more old-fashioned, cumbersome, yet acceptable, manufacturing process may have to be used. Similarly, some methods of production may be prevented upon the grounds of environmental protection. Such prevention would occur if the process created high levels of either air or noise pollution for example. Again, our roof tile firm could be affected by such controls and legislation.

(5) Input prices (*I*)

Input prices payable by the supplier cover a wide range of items as they are the prices that have to be paid for the various inputs to the good or service in question. Input prices can include: payments to factors of production such as land, capital, and labour; payments to cover the costs of machinery; and payments for any raw materials used in the production or provision of the good or service. Any rise in input prices is likely to cause the supply curve to shift to the left. The reasoning behind this shift is that for any given price less could now be produced profitably. The easiest way to understand this point is to consider the starting point of the supply curve (see Fig. 1.5): previously the producer would be willing to initiate production if the selling price of the good was equal to P_0 as this would enable the receipt of a satisfactory return over manufacturing costs at a sufficient level of output. However, a subsequent rise in costs could mean that a loss, or insufficient profit, would be made from sales at a price of P_0.

Therefore, a price greater than P_0, and one that would be high enough to cover all costs, would now be required to initiate production. Such a price could be as high as *Pz* (as seen in Fig. 1.7), and one could continue this logic at all levels of the supply curve until we arrived at a new supply curve above, and to the left of the original supply curve. This movement of the supply curve can be seen in Fig. 1.7 where the new supply curve (S_1) is created after a rise in costs. Similarly, decreases in costs could cause the

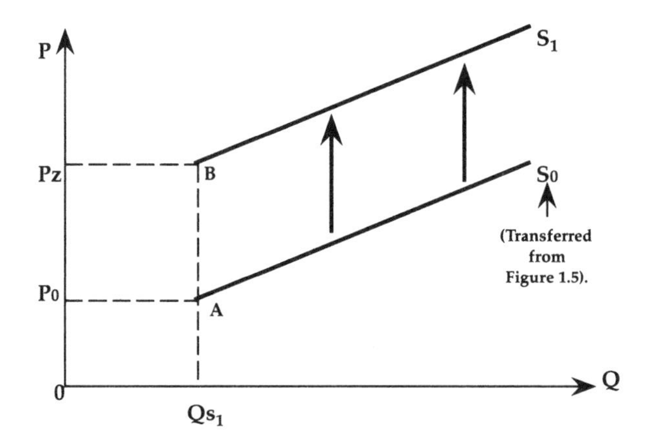

Figure 1.7. *Increasing production costs and their impact on supply*

supply curve to shift to the right. Decreases in costs could be attributable to a variety of factors such as increased mechanization and computerization, the laying off of surplus labour, increased efficiency of management and existing plant, and the lowering of raw material prices.

The total market supply curve

Just as with demand, one can find out the total supply curve of any good or service by simply adding up how much each individual supplier is willing to put on to the market at any given price. In other words the process of 'horizontal summation' is again used. If you feel it necessary imagine three universities, or three firms manufacturing roof tiles, and sum their output in order to form a total market supply curve by using hypothetical prices and quantities as was done for the analysis of demand.

Now that we have a knowledge of the fundamentals of the determinants of both demand and supply, we can bring these two sides of the market together in order to understand how the market actually works as a means of determining prices and output.

(C) The market at work: demand and supply together, and the formulation of a market equilibrium

We are now in a position to see how the price of a good or service is arrived at, and how much of it would be produced if **market forces** were allowed to operate freely by themselves. It must be recognized though that a degree of 'market failure' does occur in many markets, and consequently some form of government intervention is often justified. Therefore, a true market solution as depicted below may not always be reached. Market failure in the built environment is discussed in detail in Part 4. However, although adjustments to the market may be required with respect to equity for example, the market does provide a mechanism which should *efficiently* and *automatically* produce both a market price and equilibrium level of output for the majority of goods, services, and assets. The easiest way to understand the workings of the market mechanism is to imagine two converse situations:

1 where the initial price of a good or service was set too low;
2 where the initial price of a good or service was set too high.

These situations can be seen in Figs. 1.8 and 1.9. In both of these cases it would be a useful exercise for you to attempt to apply either of the suggested examples of the halls of residence, or the buildings materials supplier. This analysis is extremely versatile and should be able to accommodate any applied example of your choice from the built environment. Figure 1.8 demonstrates the situation whereby initial prices were set too low. If the initial price were originally set as low as P_1 few producers

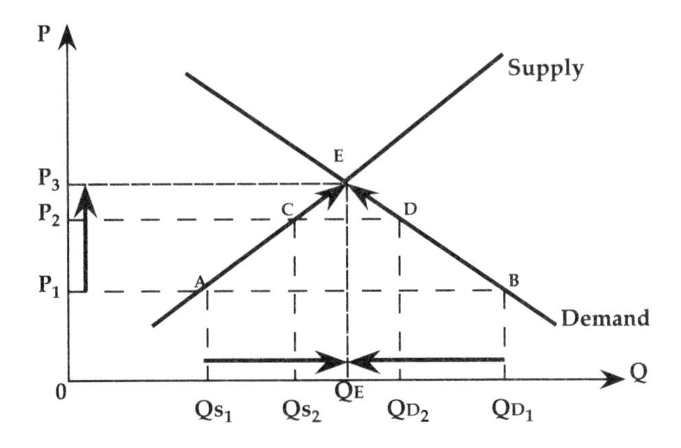

Figure 1.8. *Initial price set too low and the formulation of market equilibrium*

are likely to be willing to produce in this market as better profits could probably be made elsewhere, or few firms could cover their costs. Thus, quantity supplied on to the market will be low as represented by Q_{S1}. However, at such a low price, many people would be tempted to purchase this product as it is likely to be cheaper than most alternatives. Thus, quantity demanded is as high as Q_{D1} giving us a situation of **excess demand** equal to the distance '*AB*'. The degree of such excess demand will depend upon the relative 'elasticities' of both the supply curve and the demand curve. This point will be appreciated once you have read Chapter 2. In such cases where demand outstrips available supply, some consumers will start to offer higher prices in order to ensure that they get some of the good or service that they require. As prices are bid up in this manner, suppliers are encouraged to supply more, and thus supply increases. Simply stated, suppliers are enticed to supply more as rising prices will help enhance their profitability and it is such profitability that is the reward to entrepreneur-

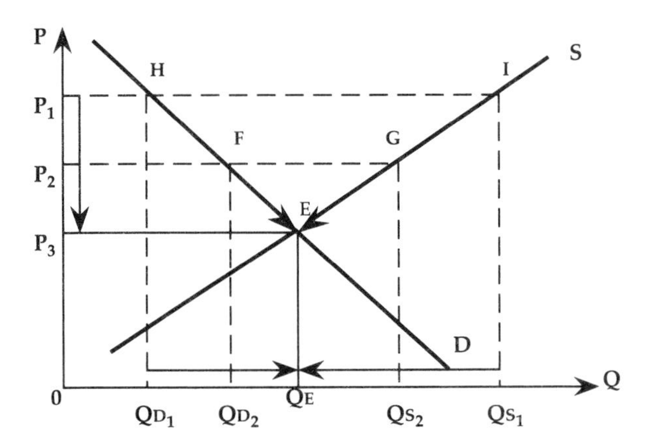

Figure 1.9. *Initial price set too high and the formulation of market equilibrium*

ship and the effort of production. As supply increases the amount of excess demand will decrease as many will not be willing to pay prices in excess of P_1. For example: as prices rise from P_1 to P_2, quantity supplied increases from Q_{S1} to Q_{S2}, yet the level of demand drops from Q_{D1} to Q_{D2}. Therefore the level of excess demand drops from '*AB*' to '*CD*' due to increasing supply and declining demand. Note that as we are dealing with changes in price here we are observing movements along both curves rather than shifts in those curves. This process will continue until we reach an *equilibrium* at *E*. With respect to the equilibrium point it is important to note that although it is efficiently and automatically arrived at in the way described immediately above, it may be viewed as being an inequitable result. That is, in some instances, the market solution may be seen as an unjust one for some people. For example, there will be many who can only afford the good or service at a price of P_1 rather than the market price of P_3. The number of people who could not afford the good at its ruling market price of P_3, rather than the lower price of P_1, is represented by the distance Q_EQ_{D1}. Moreover, one must appreciate that although this equilibrium is automatic, it is not necessarily a stable one as many variables could change over time causing a shift in demand and/or supply necessitating a new equilibrium to be formed at another point. The speed at which an equilibrium will be re-established will depend upon the efficiency of the market in question, and whether or not there are any artificial constraints on that market.

Using an identical logic one can see the sequence of events that lead to the formulation of an equilibrium when prices are initially set too high. This mechanism is shown in Fig. 1.9. Here, if suppliers originally charge too much by setting prices too high at P_1, they would be encouraged to supply a large amount of their product on to the market (Q_{S1}) as they would be anticipating a high return on their sales due to the high price. However, at such high prices, few could afford the product and demand would be low at Q_{D1}. Thus, we would have a situation of **excess supply** equal to the distance *HI*. In an attempt to avoid the unnecessary and costly accumulation of stocks, or possible wastage, producers are likely to react to this situation by lowering prices. Such price reductions will help to stimulate more demand as the product becomes more affordable, and reduce supply as producers realize that their expectations of price were too optimistic. For example, as price is reduced to P_2, the amount of excess supply has been reduced from *HI* to *FG*. This process will continue until again we reach a position where the desires of producers and consumers are the same, or expressed in another way: the market is in equilibrium. A useful exercise for you to attempt now is to shift either the demand curve or the supply curve by changing one, or more, of the explanatory variables, and work through the logical sequence of events of how a new equilibrium point would be established.

An overview of equilibrium: a return to the example of halls of residence

From the above analysis it can be seen that if the university were to set hall rents too low there would be a large number of students seeking such accommodation as it now became relatively cheap in comparison with other forms of rented property. At such low rents, however, the university may find it difficult to cover the operating costs of such accommodation and would therefore be unwilling to supply (build) more halls of residence. Due to the level of excess demand many students will be unable to obtain a room in hall, and therefore in an effort to secure a room they may indicate a willingness, via student representatives, to pay higher rents. As rents are bid up in this way the university has more of a financial incentive to provide more of such accommodation, although the level of demand will begin to drop as such rooms attract a rent that is too high for some to pay. As we have seen above, it is in this manner that an equilibrium could be reached. Conversely. if the university were to set its rents too high it may find that few students were attracted to their properties so that a situation of excess supply would exist. The only way that this excess supply could be reduced is by a lowering of rents so as to attract more students away from renting alternative forms of accommodation.

(D) The market mechanism at work: some shifts in demand and supply

One could have chosen a whole range of potential shifts in demand and supply in order to illustrate the impact of such movements in the market, however so as not to labour the point I have selected two examples which should be sufficient to make the issue clear:

(a) The impact of declining demand on the market

If the demand curve were to shift to the left for a good or service one would eventually see the formation of a new market equilibrium at both a lower price and a lower level of output. As mentioned earlier, the speed of such an adjustment to a new equilibrium point would depend upon the product in question and the efficiency of its associated market. In Fig. 1.10 a drop in demand from D_1 to D_2 has caused a drop in price from P_1 to P_2 and a fall in the quantity traded of the product from Q_1 to Q_2. Therefore in the case of the halls of residence, if the demand for student accommodation were to drop because of a demographic decline in student numbers for example, the university may find that the only way that it can maintain its relative 'market share' in the rented student accommodation market, and to avoid vacant rooms, would be to lower rents. With reduced levels of demand the university would only be able to let Q_2 rooms at a price of P_2. However, if they had kept rents as high as P_1, Q_2Q_3 students would have found alternative, presumably cheaper accommodation. Likewise, if the

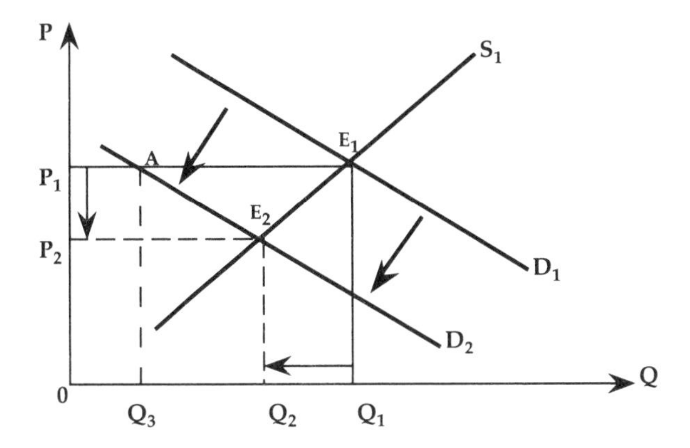

Figure 1.10. *The impact of declining demand on the market*

roof tile manufacturer were to face declining demand due to a recession in the building industry he would be wise to 'follow the market' and reduce his prices, as maintaining prices at 'pre-recession' levels could lead to an even greater loss of orders. It is for this reason that the more inefficient, high cost, firms go out of business in recessionary times of declining demand.

(b) The impact of a reduction in supply on the market

A reduction in supply would cause the supply curve to shift to the left as less can be supplied at any given price. Such a shift in the supply curve can be seen in Fig. 1.11 where the supply curve has shifted from S_1 to S_2 leading to a drop in the quantity traded on the market from Q_1 to Q_2, and an increase in price from P_1 to P_2 reflecting the new, relative scarcity of the

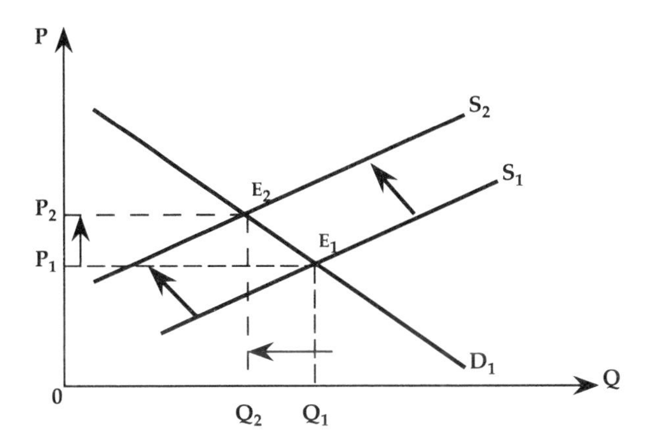

Figure 1.11. *The impact of a reduction in supply in the market*

item in question. For example, if one of the university's halls of residence had to be closed down for extensive repairs, or renovation work, this could lead to a shortage of such accommodation that could drive rents higher as students competed for the limited number of rooms available. Likewise, the roof tile manufacturer may suffer a technical problem in a production plant that forces a drop in output. As fewer roof tiles are now being supplied on to the market their price will rise as builders compete for the limited supply. Again both of these results will depend upon the relative elasticities of both the demand and supply curves. Therefore, in order to understand the market mechanism in fuller detail this book will now go on to consider 'Elasticity'. However, before progressing to the next stage it may be a useful exercise for you to now draw examples of *increasing* demand, and *increasing* supply, and then follow through the logic of price and quantity changes. Selecting examples from the built environment will help you to appreciate the application of the theory as well as promoting an understanding of it.

— 2 —

Developing market analysis for further, and more precise, application

Obviously as students of 'Built Environment' courses your main interests here should lie in applied examples and analysis. Therefore, in the penultimate part of this section (Chapter 3) the book will concentrate on construction and property orientated case studies in some detail. However, to fully appreciate these examples, and to understand the versatility and application of market theory, we will now briefly consider the concept of 'elasticity'.

Elasticity

Elasticity is a tool of measurement that gives us far more precision with respect to our market analysis. It is a very important 'tool' at our disposal, yet it is fortunately very easy to understand as although there are three types of elasticity to consider, only one formula (with minor suffix adjustments) and two key principles need to be learnt. The three categories of elasticity are:

1 price elasticity;
2 income elasticity;
3 cross–price elasticity.

We will now go through each of these forms of elasticity in turn, although chiefly concentrating upon price elasticity as it is price elasticity that is one of the most commonly used measures in property market analysis.

(A) Price elasticity of demand (*Epd*), and price elasticity of supply (*Eps*)

Price elasticity (*Ep*) measures the responsiveness of the quantity demanded, or the quantity supplied, due to a change in price. More specifically it tells us exactly how much quantities will change for each one per cent change in price. You will find that by looking at other textbooks that there are a variety of formulae available to you in order to work out price elasticity. These formulae range from those giving you **point elasticity** (the elasticity at any given point on a curve), to those giving you **arc elasticity** (the elasticity between any two given points on a curve). Although the 'mathematical make up' of these formulae is often different, they should obviously provide you with the same, consistent answers. However, many of them are unnecessarily complex and do not always work in both directions. That is, they give you a different answer for price increases than they do for decreases in price. Thus, although we could debate the relative merits and weaknesses of these formulae, such arguments and theorizing are *not* what this book is about. Therefore for the purposes of straightforward and simple application, a general formula can be used, and as mentioned earlier, the following formula can be used for all categories of elasticity with only minor adjustments:

$$Ep = \frac{dQ}{\left(\frac{Q_1 + Q_2}{2}\right)} \div \frac{dP}{\left(\frac{P_1 + P_2}{2}\right)}$$

where Ep = price elasticity;
d = change in
Q = quantity;
P = price.

Now, to fully understand both the workings of the formula and the meaning of its results it is best to refer to the text in conjunction with Fig. 2.1. In the case of demand we would use the formula in the following way (bearing in mind the annotation given in Fig. 2.1a):

$$Epd = \frac{dQd}{\left(\frac{Qd_1 + Qd_2}{2}\right)} \div \frac{dP}{\left(\frac{P_1 + P_2}{2}\right)} = ?$$

Where: Epd = price elasticity of demand;
dQd = the change in quantity demanded;
Qd_1 = the initial quantity demanded;
Qd_2 = the second quantity demanded after a change in price;
P_1 = the initial price;
P_2 = the new price.

With demand we will find that all of our answers to this elasticity formula will be *negative*. However, do not be alarmed or confused by this, as it is simply a result of the fact that we have a negative, or inverse, relationship between price and quantity demanded. That is, when price goes up, quantity demanded falls; and when price falls, quantity demanded increases. Hence, the reason why we observe a negatively, or downwards, sloped demand curve in diagrammatic form. As all the answers to the equation are negative, and due to the fact that we normally think in terms of positive values, it is quite normal, or in fact convention, to disregard the negative sign altogether, or at least put it in brackets. For example, if when examining the demand for a particular good we found that our answer to the above equation was:

$$Epd = (-)0.6$$

it tells us, as a guide, that between points *A* and *B* on the demand curve that we are considering (see Fig. 2.1a), every one per cent change in price will only lead to a 0.6 per cent change in quantity demanded. For example, for every one per cent *increase* in price there will be a corresponding 0.6 per cent *drop* in the level of demand.

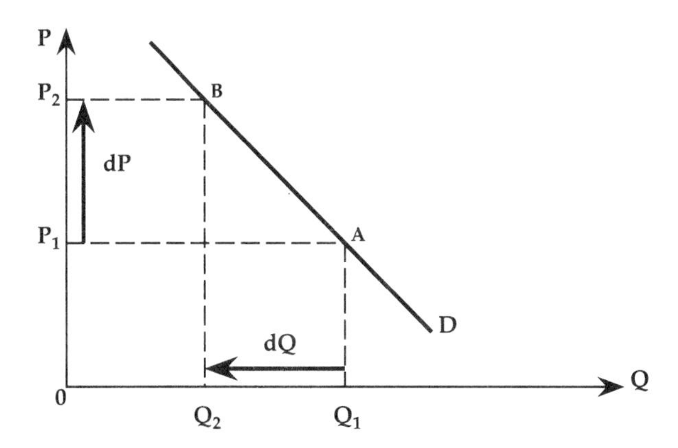

Figure 2.1a. *Price elasticity of demand*

With respect to the elasticity of supply, *exactly* the same principles are used. The only alterations that have to be made are that we now simply insert our supply curve values into the equation. This marginally amends the formula so that it reads:

$$Eps = \frac{dQs}{\left(\frac{Qs_1 + Qs_2}{2}\right)} \div \frac{dP}{\left(\frac{P_1 + P_2}{2}\right)} = ?$$

Where Eps = price elasticity of supply;
dQs = the change in quantity supplied;
Qs_1 = the initial quantity supplied on to the market;
Qs_2 = the new quantity supplied after the change in price;
P_1 = the initial price;
P_2 = the new price.

With supply we will find that all of our answers to this formula are *positive* values. This reflects the fact that we have a positive relationship between price and quantity supplied. That is, when price goes up, the incentive to supply more increases. Hence we observe a positively, or upwards, sloping supply function when viewed in diagrammatic form. For example, if our answer to the above equation depicting the price elasticity of supply was:

$$Eps = 2$$

it tells us, as a guide, that between points *C* and *D* on the supply curve in Fig. 2.1b, every one per cent change in price will cause a two per cent change in the supply of the particular good, service, or asset in question. For example, a one per cent rise in prices should encourage the suppliers in this case to increase output by two per cent.

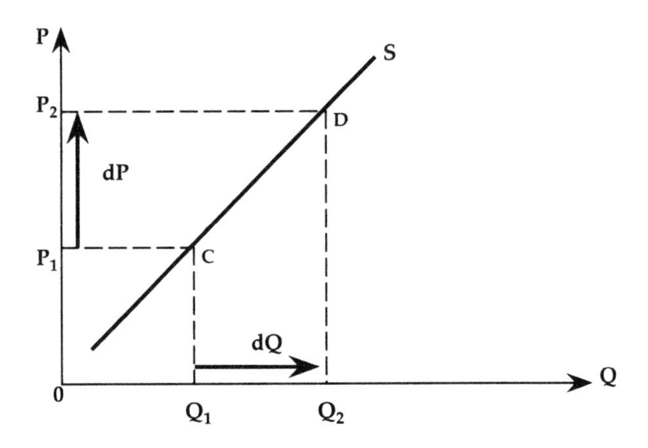

Figure 2.1b. *Price elasticity of supply*

In relation to both demand and supply it should be realized that, apart from a few exceptions, the precise value of price elasticity will vary at different points of the curve. However, although it can be dangerous to generalize, the results will enable one to predict with a greater degree of accuracy how markets will react to changing circumstances. For example, a knowledge of elasticity should enable one to predict how much house prices in any given area would change due to a known, rising demand, simply by knowing the current and expected elasticity of the supply of housing in that area, (see Chapter 3A). In order to fully understand how this can be achieved you should consider the following two categories of potential results that you may obtain from your equations:

1 Relatively *elastic* demand or supply;
2 Relatively *inelastic* demand or supply.

Although there are certain exceptions and extremes to these two categories the exceptions are rare, and therefore these two general results are the most common. The book will now deal briefly with each in turn:

(1) Relatively elastic demand or supply

The situation of a 'relatively elastic' demand or supply curve exists where *a change in price causes a more than proportionate change in quantity demanded or quantity supplied*. For example, in the case of demand, a small increase in price will cause a more than proportionate decline in quantity demanded. That is the consumption of that good or service is quite substantially reduced because of a relatively small increase in price. Likewise with supply, a small increase in price, for example, will encourage much greater output on to the market. In other words, the response of the supplier is more than proportionate to the change in price. In either the case of supply or demand, a relatively elastic situation will give rise to the answers of the price elasticity formula being *greater* than one. That is:

$$Epd = >(-)1$$

and:

$$Eps = >(+)1$$

In other words, a one per cent change in price creates a more than one per cent change in quantity demanded or quantity supplied. Assuming equally graded axes on our graphs the notion of a relatively elastic result is probably best envisaged diagrammatically by visualizing relatively *flat* demand and supply curves with gentle gradients as seen in Figs. 2.2a and 2.2b. In both the graphs illustrated in Figs. 2.2a and 2.2b one can see that due to the low gradients of the curves in each case relatively small changes in price from P_1 to P_2 have led to correspondingly larger changes in quantity as represented by the distances Q_1Q_2 in each instance. Thus, for example, if our calculations of a demand curve arrived at a figure of:

$$Epd = (-)3$$

it implies that for every one per cent increase in price there is a three per cent reduction in quantity demanded. One explanation of this outcome

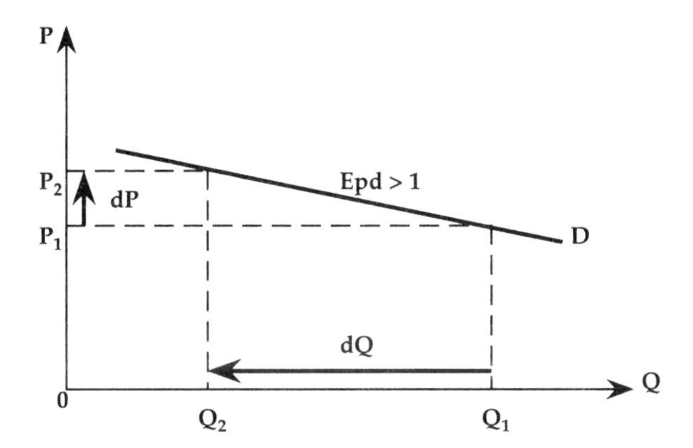

Figure 2.2a. *Relatively elastic demand*

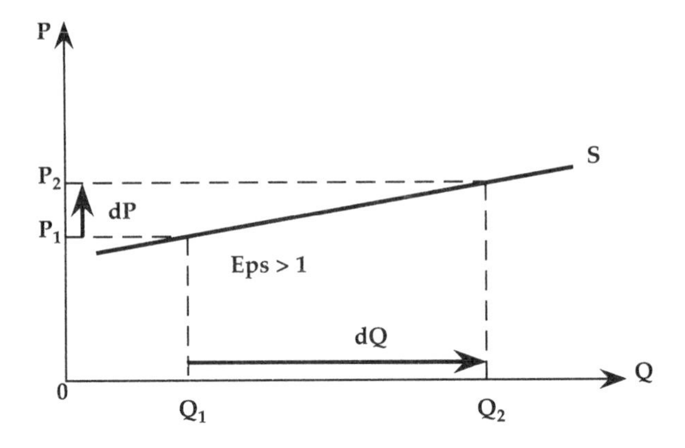

Figure 2.2b. *Relatively elastic supply*

could be that consumers can easily switch into a substitute good if the original good becomes too expensive. This part of the section will shortly deal with the determinants of elasticity in more detail. Similarly, if we were to find that our answer to the equation calculating the elasticity of supply was, for example;

$$Eps = 5$$

it implies that every one per cent increase in price encourages a five per cent increase in the amount supplied on to the market. For example, in this case, if the demand for the product increased thus causing such an increase in price, it would seem that suppliers could easily increase output, perhaps by being able to utilize and employ more unskilled staff, by making use of 'night shifts', or by enhancing productivity via better management.

The unlikely extremes of this situation would be a **perfectly elastic**, or, in other words, a **completely horizontal** demand or supply curve with no gradient at all. For example, Fig. 2.3 illustrates a perfectly elastic, horizon-

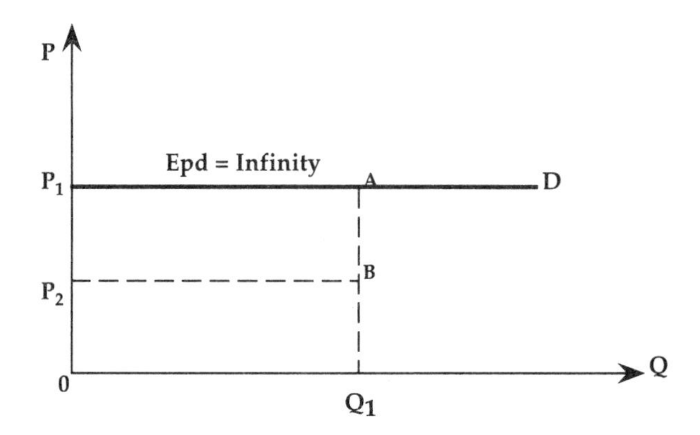

Figure 2.3. *Perfectly elastic demand*

tal, demand curve. If a firm were to be faced by such a demand curve it would imply that if the firm attempted to raise the price of its product above P_1 it would sell nothing at all as there is no demand for this product at any price in excess of P_1. This outcome may be due to the fact that consumers of the product could purchase a perfect substitute for the original product also at a price of P_1, and therefore there would be no point in paying anything in excess of this price in order to obtain it. On the other hand, if the firm charged less than P_1, say P_2, it would not be maximizing its potential revenue. For example, more revenue would be received from selling an output of Q_1 at a price of P_1, rather than at the lower price of P_2. Note that revenue is simply the amount sold multiplied by its selling price, $(P \times Q)$. Therefore the revenue received from selling Q_1 at P_1 is given by the area of the rectangle $0P_1AQ_1$ which is obviously greater than the rectangle $0P_2BQ_1$ that would be received by selling Q_1 at a price of P_2. Although such an extreme is an unlikely occurrence, it is a useful comparative analytical point with which to examine either demand or supply curves that are approaching or have approached this situation. In fact, use of the perfectly elastic demand curve is made when one examines the behaviour of firms operating in conditions of very high or 'perfect' competition where the price elasticity of demand is potentially equal to infinity. This analysis can be seen in Chapter 7.

In conclusion to the debate on relatively elastic curves, it can be said that the higher the value of the result of the equation, the more horizontal will be the curve, and therefore the greater the degree of elasticity.

(2) Relatively inelastic demand and supply

A **relatively inelastic** demand or supply curve essentially exhibits the reverse characteristics of their elastic forms. With a relatively inelastic curve *a change in price will cause a less than proportionate change in quantity demanded or supplied.* For example: an *increase* in price will only cause a small drop in quantity demanded in the case of demand, and only a small increase in quantity supplied in terms of supply. Therefore, with all curves that exhibit 'inelasticity', the answer to our elasticity equation will be *less* than one. In other words:

$$Epd = <(-)1$$

and:

$$Eps = <(+)1.$$

That is, for every one per cent change in price there will be a less than one per cent change in quantity demanded or supplied. Again, assuming uniform axes, this is perhaps best understood diagrammatically where it can be seen that inelastic curves will have steep gradients, and the more inelastic they are the steeper their gradient will be. Both inelastic supply and demand curves can be seen in Figs. 2.4a and 2.4b. Here it can be seen that relatively large increases in price from P_1 to P_2 only induces relatively small changes in quantity from Q_1 to Q_2. Thus, for example, if by using our equation we arrived at an answer of :

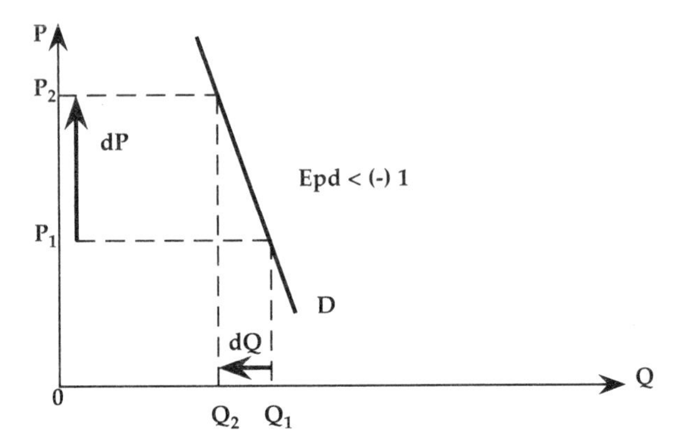

Figure 2.4a. *Relatively inelastic demand*

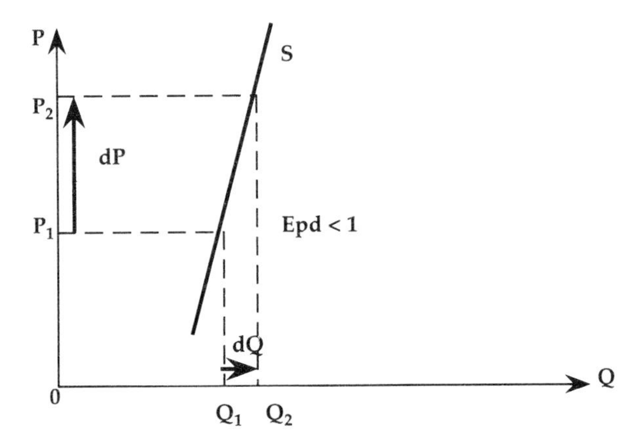

Figure 2.4b. *Relatively inelastic supply*

$$Epd = (-)0.5$$

it would imply that for each one per cent increase in price, quantity demanded would only drop by one-half of one per cent. This response could perhaps be due to the fact that the good is viewed as a near necessity of life and as such consumption cannot be reduced too radically in the light of increasing prices. The more inelastic the curve the lower this value would be. Likewise, if when calculating the elasticity of a supply curve we found that the answer to our equation was of the form:

$$Eps = (+)0.2$$

it implies that for every one per cent increase in price, say due to a rise in demand, will only lead to an increase in supply of one-fifth of one per cent, at least in the short run. Such an inelastic result could indicate an example of a good that is difficult or time consuming to produce, and therefore changes in demand cannot be reacted to quickly. As with demand, the lower the figure given by the equation the more inelastic will be the curve. Factors determining the degrees of such elasticity and inelasticity will be examined shortly. Yet again, the extreme case of inelasticity, that is a **perfectly inelastic** curve, is unlikely to occur in reality, but is shown for purposes of comparative analysis in Fig. 2.5. Figure 2.5 shows a perfectly inelastic supply curve whereby any change in price causes no supply response whatsoever. As quantity does not change the answer to our equation will be:

$$Eps = 0.$$

The book will examine the likelihood of such a result in Chapter 3B when the case of building land is considered.

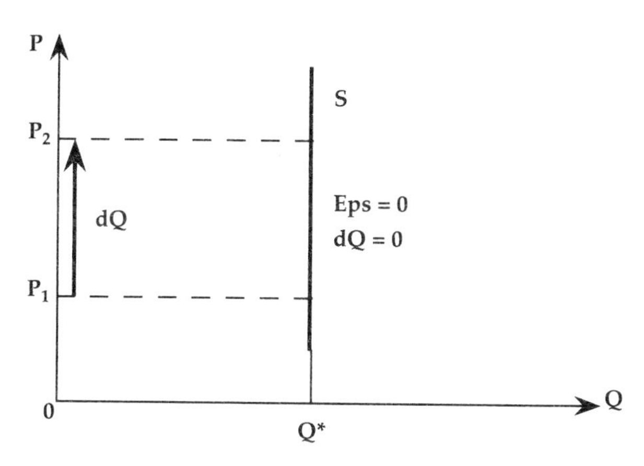

Figure 2.5. *Perfectly inelastic supply*

A conclusion on the measurement of price elasticity

The 'spectrum' of possible results from our elasticity equation, ranging from the case of perfect elasticity to the case of perfect inelasticity, is shown in Fig. 2.6. When examining this spectrum it is important to remember that, in reality, one generally gets relatively inelastic or relatively elastic results, although it must be appreciated that the extremes can be useful points of comparison. Moreover, it must be realized that the actual figure for elasticity will vary at different points along any particular curve. For example one point on the curve will exhibit 'unitary elasticity', that is where elasticity is equal to one, but for the whole curve to exhibit unitary elasticity it would need to be a rectangular hyperbola. That is for each one per cent change in price there would be a one per cent change in quantity. Again, this represents an unlikely result, but is a useful comparative measure as unity represents the dividing line between elasticity and inelasticity. Although a

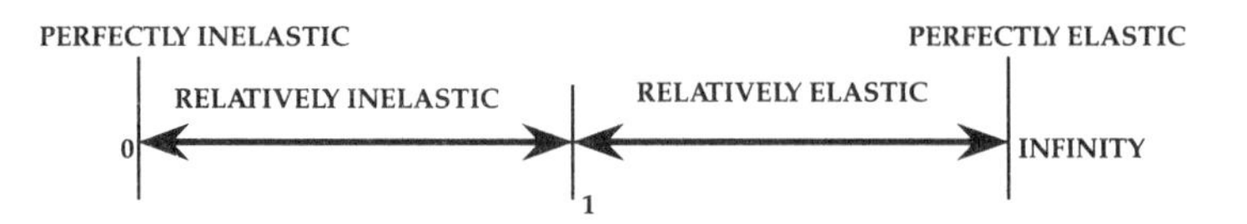

Figure 2.6. *The elasticity spectrum*

formula has been suggested for you to use, it is often not necessary for you to know the exact figure for the inelasticity or elasticity in question as just a knowledge of whether a curve exhibits either result should enable you to conduct the correct analysis and give relatively accurate advice. Having understood the actual measurement of price elasticity, this chapter will now go on to consider what factors actually influence the degree of elasticity or inelasticity of demand or supply.

Factors influencing the price elasticity of demand

The following are perhaps the key determinants of elasticity, although they should not be considered as an exhaustive list. Moreover, these 'determinants' are not mutually exclusive from one another as more than one can exert an influence at any given time.

(a) The degree of substitutability and complementarity

Obviously goods, services, or assets which have **substitutes** tend to be relatively **price elastic in demand** as per the demand curve shown in Fig. 2.2a. In fact, the higher the degree of substitutability between the items under examination, the greater the degree of elasticity. In such circumstances the quantity demanded for the original item becomes relatively sensitive to a change in price as consumers can easily use an alternative if the price of the original item increases. If there existed a perfect substitute for the original item one would observe a perfectly elastic demand curve as seen and discussed in Fig. 2.3. Whereas a lack of substitutes tends to give rise to a relatively inelastic demand curve, as seen in Fig. 2.4a. In the case of an inelastic demand curve, even if price rises for the original item, consumption cannot be varied greatly as few, or no, alternatives can be found. The actual result will also depend upon the 'luxury/necessity syndrome' as discussed in point (b) immediately below. With respect to goods, services, or assets that are seen as **complements** to one another, the degree of elasticity or inelasticity will depend upon whether the original good is inelastic or elastic in demand. For example, if the original good is viewed as being inelastic in demand, roughly the same amount of it will be consumed after a price change as before the price change, therefore any items that are consumed with it will also only react with small quantity changes. Therefore, inelasticity of demand or elasticity of demand of the original good will correspondingly give us inelasticity and elasticity of demand for the associated complementary good. For example, bricks and mortar can be viewed as complementary goods to one another as one is of little use without the other. If the price of bricks were to increase most builders would have to bear the brunt of the price increase as they would still need to buy bricks in order to construct their buildings. In other words, their demand for bricks is relatively price inelastic. Their demand curve for bricks would only become more elastic if they could change their reliance on bricks by either changing the manner in which they built buildings, or changed the design of the building, or were to replace brickwork with an alternative material. Many of these options may not be acceptable from the point of view of the client for which the building is being built. Therefore, as bricks are still being used in roughly the same quantities as before the price rise, roughly the same quantities of mortar will also be used. Thus an inelasticity in the demand for bricks has led to an inelasticity in the demand for mortar.

(b) The 'luxury/necessity' syndrome

Logically **necessities** will be characterized by **relatively inelastic demand curves** as seen by the demand curve shown in Figs. 2.4a and 2.4b. In fact, the more essential the item, the higher the degree of inelasticity that would be observed. For example, in the case of increases in price for essential goods, quantity demanded would hardly alter. Thus, it is likely that the demand for 'housing', for example, is highly inelastic as we will all seek some form of shelter (see Chapter 3A). If such goods do increase in price, yet consumers still have to purchase them, this is normally made possible by the consumer reducing some other form of non essential purchases, or obtaining a loan to cover the increased costs. On the other hand, if a good or service is perceived as a luxury by consumers its demand is likely to be **highly elastic**. Such a highly elastic demand curve is of the form shown by the demand curve in Fig. 2.2a. Therefore, in such cases, if the item is not an 'essential of life', and if it did go up in price, consumers would be able to cut back on its consumption. For example, if the cost of non essential extensions to one's house, such as the building of conservatories, were to increase, the number of people demanding such work could easily drop without impairing the existing utility derived from the house in its present form. This result would not be possible if the price of an essential service such as electricity or water were to increase as these need to be purchased in consistent quantities to maintain the existing utility derived from the house. If electricity consumption were to be reduced, for example, the occupier may have to suffer a decline in heating, or quantities of hot water.

(c) The proportion of income spent upon a good or service

The less that you spend on any particular good or service, the less likely that your demand would respond significantly to a change in price. Therefore, such a situation will lead to a relatively inelastic demand curve (see Fig. 2.4a). For example, if a building firm is faced with an increase in the cost of bathroom tiles from its supplier, the firm may not cut its demand down substantially as such an increase represents a relatively small proportion of total building costs. Moreover, bathrooms will still need to be tiled although savings could be made by putting fewer rows of tiles in each bathroom.

(d) Broadness of category

When undertaking any research and analysis, it is of paramount importance to carefully define the 'broadness of category' that you are dealing with so as to avoid examining highly misleading results. To demonstrate this point consider the following example: If the research area is defined *very broadly* such as residential accommodation in general, we are likely to arrive at a highly inelastic demand curve (see Fig. 2.4a) as irrespective of price people will have to obtain shelter somehow. However, if we are looking at a single type of housing (for example new houses with mock-tudor frontages) demand will be highly elastic as there are so many substitutes available (see Fig. 2.2a).

(e) The time period

The time period is a most important variable to appreciate as elasticity will tend to change over time. Specifically people's reactions continue to change in some cases well after the initial price change. Or, with respect to gradual price changes, the longer it takes for the price change to occur, the easier it becomes to find a substitute if necessary, thus the more elastic the demand curve will become (see Fig. 2.2a). For example, as petrol prices rise, people will still need to use their car and are unlikely to be able to alter their demand radically in the short run. However, in the medium term, they could buy a smaller engined, more economical, vehicle, switch to a diesel car, make fewer non-essential trips and so on. Thus, in the medium term the demand for petrol may well become more elastic despite an initial high degree of inelasticity. Despite this demand behaviour, history demonstrates that even very large price increases are eventually forgotten in the long-run as people become accustomed to the general higher level of prices. In such instances, long run consumption behaviour tends to return to the inelastic phase once again. This is an important conclusion for the town planner concerned with the provision of roads and car parking facilities in the urban area as medium term decreases in the use of private transport do not necessarily mean a long-run reduction in the need for improvements in this field.

Factors influencing supply elasticity

A variety of factors will determine the elasticity of supply, however the two most important determinants are perhaps: the ease of response, and the time factor. Both of these issues are now briefly discussed.

(a) Ease of response

The ease at which producers can adapt their equipment and production processes is a key factor to contributing to the degree of supply inelasticity or elasticity. The more straightforward this is, the easier it becomes to adjust output in response to any price changes, and thus such ease tends to be reflected in a **relatively elastic** supply curve (see Fig. 2.2b). Conversely, an inelastic supply curve suggests that any increase in demand will cause large increases in price yet little change in output. Such supply inelasticity will occur if the producer finds it difficult, or inappropriate, to increase output (see Fig. 2.4b). For example, this situation can arise when there are difficulties in obtaining a particular input to the production process, or the necessary finance, to be able to undertake additional production. Thus, a brick manufacturer is likely to be able to increase supply with ease in response to an increase in demand because, subject to the availability of raw materials, the production process could be used to greater capacity in most cases. However, on the other hand, the house builder (see Chapter 3A) will not suddenly be able to increase output in response to increases in the demand for housing, as suitable development sites need to be found, planning permission sought, and so on. Therefore, the brick manufacturing process is likely to be characterized by a relatively elastic supply curve (see Fig. 2.2b), whereas the supply of housing is likely to be highly inelastic (see Fig. 2.4b).

(b) The time factor

The longer the available period of time the easier it becomes to obtain the resources needed for increased output. Therefore, the longer the time period, the supply curve is likely to become more elastic. For example, after initial price hikes caused by rapidly rising demand, producers cannot often react sufficiently quickly, but after time they can begin to make a response to the new market situation. In the early stage of a housing boom where there is a rapidly rising demand for owner occupied housing, 'reactive builders' (rather than speculative builders) may be caught unawares and will need to find building land, planning permission, etc., before they can increase output in the form of building more houses. Thus, it is only in the long run that their supply curve becomes more elastic. However, care should be taken by the house builders not to over-react to any increases in demand by obtaining too much land, or by building too many houses, as these may not be traded profitably if the market subsequently slumps. In this situation the market would be characterized by excess supply until house prices dropped in order to clear the market.

(B) Income elasticity of demand

Income elasticity of demand is basically the same concept as price elasticity of demand, except that now, with income elasticity, one can examine the responsiveness of quantity demanded due to a *change in income* rather than a change in price. Thus we can adjust our basic elasticity formula to read:

$$Ey = \frac{dQd}{\left(\frac{Q_1 + Q_2}{2}\right)} \div \frac{dY}{\left(\frac{Y_1 + Y_2}{2}\right)} = ?$$

where: Ey = income elasticity;
dy = change in income;
Y_1 = original income;
Y_2 = new income.

The resulting answer to this formula is usually a **positive figure** as most goods are 'normal goods' where there is a direct relationship between quantity demanded and income. For example, if your income increases you tend to buy more normal goods, thus quantity demanded increases. If one obtains a **negative figure** from the formula it implies that the good is an 'inferior good'. In the case of inferior goods, people tend to buy less of them as their income increases. An example of an inferior good that would exhibit such a response is cheap home fittings such as self-assembly furniture. Cheap forms of construction, small buildings, or buildings in poor locations, may also be viewed as inferior goods. For example, whether one considers the house buyer, or the commercial client seeking suitable business premises, both parties may be forced to purchase inferior, low grade dwellings in times of low income, or an economic recession, but would hope to move into superior premises once the economic climate has improved.

The relationship between income and quantity demanded can be seen by referring to Figs. 2.7a and 2.7b, whereby a positive gradient represents a normal good, and a negative gradient represents an inferior good.

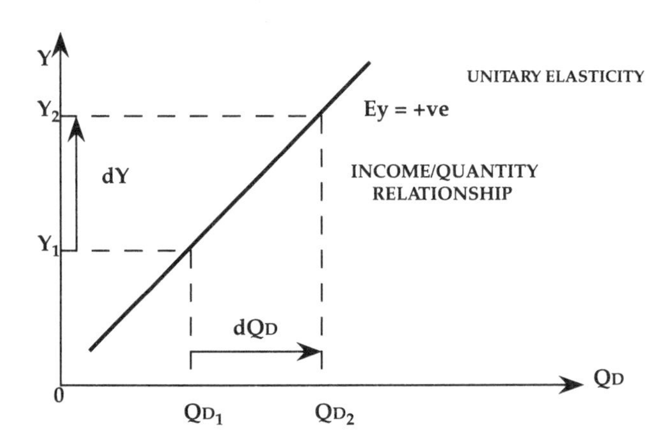

Figure 2.7a. *A 'normal' good: a positive gradient*

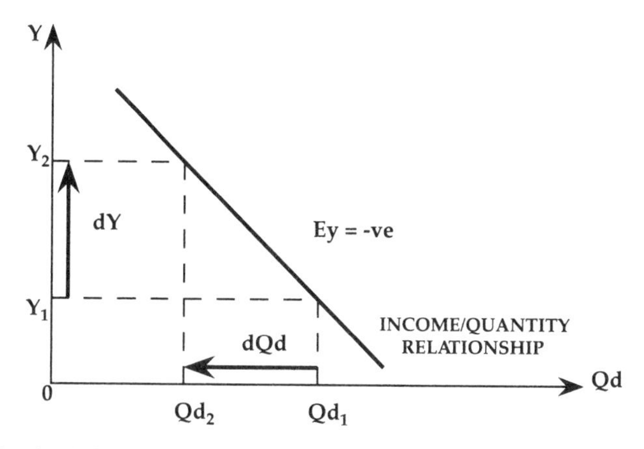

Figure 2.7b. *An 'inferior' good: a negative gradient*

(C) Cross elasticity of demand

Yet again, the idea of cross elasticity of demand is not significantly different to the initial concept of price elasticity, except that this time one is examining the effect on demand for a particular good or service in relation to a change in price of *another* related good or service. This measurement enables one to assess whether the goods or services that we are examining are substitutes or complements, and the degree to which they fall into these two categories. The formula for cross price elasticity of demand can be written as:

$$Ex = \frac{dQa}{\left(\frac{Qa_1 + Qa_2}{2}\right)} \div \frac{dPb}{\left(\frac{Pb_1 + Pb_2}{2}\right)} = ?$$

where Ex = Cross elasticity of demand.
a = good or service 'a';
b = good or service 'b'.

For example: if two building materials or components were used in conjunction with one another they would be seen as complements. Thus, if a particular brand of metal window frame required a specific type of fixing to secure it to a building, these two goods would be viewed as being complementary. If the fixings were to become too expensive, one is likely to see a decline in the demand for such fixings and thus a corresponding decline in the demand for metal window frames – an alternative, such as plastic or wooden window frames, would be used as they would now be cheaper to install in total. The strength of this complementarity would depend upon whether these fixings were the *only* feasible securement method, or whether there were alternative methods available. Therefore, in all cases exhibiting complementarity one will obtain a *negative* number from the formula as the relationship between the price of the complement and the demand for the original one is *inverse*. Thus, the more the goods are seen as complements, the greater the negative value of the formula. On the other hand, the formula could indicate that two goods were perceived as substitutes – such as metal and wooden window frames. In the case of substitutes one would expect a positive relationship between movements in the price of one good and the quantity demanded of the other, and thus a *positive* answer to the formula. For example, if the price of metal window frames were to *increase* the quantity demanded of wooden window frames is likely to *increase* as more will be used as they become relatively cheap. Thus, the more positive the answer to the formula, the more the goods are seen as substitutes.

The above definitions and analysis have stressed that market analysis can be made more detailed and accurate by looking at the responsiveness of the market to changes in price, and income, and the behaviour of other related goods. Once these points have been appreciated we can now move on to examine case studies related to the built environment in detail (see Chapter 3).

— 3 —

Land and construction related case studies

The following part of the text consists of some brief construction and land related case studies using market analysis. These case studies are by no means fully exhaustive as one could easily write a book on each. However, they are intended to be relatively thorough starting points for discussion especially when examined in conjunction with up-to-date data. No numerical data has been used in this part as it is felt that such data would easily date and tie the analysis down to a specific country or geographical area. Importantly though, the principles discussed here should largely hold irrespective of the date that you can apply to them, and such application is left up to you. Specifically, it will be a useful exercise for you to seek out the various key sources of data that are available in order to add some empirical detail to the theory.

3A The housing market

The housing market is often a popular area of study and research, perhaps primarily due to the fact that as most of us live in houses of one form or other, (whether we live in public sector (council) accommodation, the private rented sector, or are owner occupiers) we can easily identify with the good in question. Moreover, an awareness of the housing market is especially keen in countries which have a very high level of owner occupation, coupled with relatively high incomes. It is in such countries that housing is not just considered as somewhere to live, but as an investment asset also. Furthermore, because of the general level of public interest in this market, the media often selects housing as a key issue of debate, rather than the commercial and industrial property markets which are, perhaps wrongly, seen as being more removed from people's direct interests. This is especially true considering that the demand for the latter two types of property is a derived one, and as such the link with the general public is not as obvious.

The analysis which now follows is concerned with the *owner occupier housing market*. We could initially look at this housing market in a relatively simplistic manner by using the elementary theory and observations made in the earlier parts of this section. With such a knowledge we would be able to ascertain the relative importance of the variables seen in our initial demand and supply functions. Namely:

$$D = f(P,Y,S,T)$$

$$S = f(P,A,S,T,I).$$

However, with housing, as with most other construction markets, we need to look at a much broader spectrum of issues and variables. To appreciate this point let us now carefully consider the supply of housing, and then go on to discuss demand, in order to see how the market actually operates.

The supply of housing is primarily inelastic, at least in the short run. That is, irrespective of large increases in demand, few new houses are supplied on to the market. This is despite the fact that it would normally be in the interests of house builders to supply more housing as demand increases so as to reap the potential profits from a buoyant market. Such supply inelasticity is shown in Fig. 3.1. Here it can be seen that such inelasticity leads to large changes in price as a response to a demand change, yet small changes in the quantity traded. Specifically, an increase in demand from D_1 to D_2 has led to an increase in house prices from P_1 to P_2, yet the volume of houses traded on the market has only increased to Q_2 from Q_1. The reasons for this supply inelasticity are twofold:

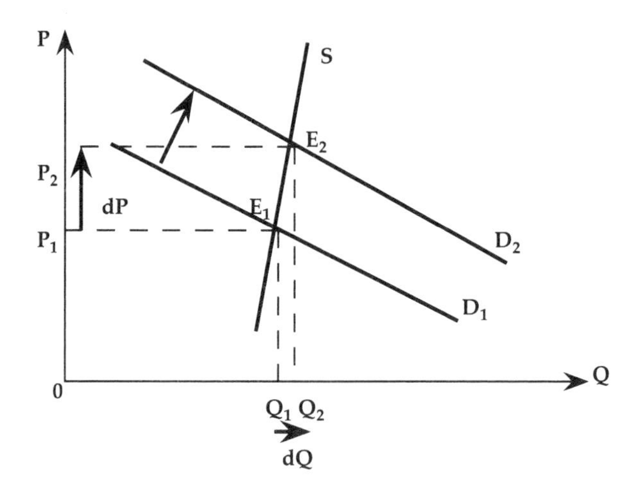

Figure 3.1. *The supply of housing in the short run*

1 If most building firms are reacting to the market, rather than basing decisions on forecasts of change, as in the case of the speculative builder, it will take time for them to realize that demand is rising in the first instance. Even after recognizing increases in demand, they will need to acquire the necessary building land (in the absence of a 'land bank'), plan the development, seek planning approval, organize production and so on. In fact, completion is likely to take place in phases, the first phase 'coming on line' a good year or so after project initiation. Thus, our first

problem is that of a considerable **time lag** in the supply process of getting new output (houses) on to the market.

2 It must be realized that, in most countries, the existing housing **stock** is so large in the first place, that any annual **flows**, or additions, to this stock are unlikely to be that significant in relation to the overall number of houses. Typically additions to stock are usually represented by increases in the order of one or two per cent per annum. In other words the existing, 'second hand' properties dominate the market.

Despite this inelasticity however, new building does of course occur, and other additions to stock can come from a variety of other sources such as transfers from the rented sector, sales of public sector dwellings, household dissolution and emigration. Moreover, builders can improve their performance with respect to supply by attempting to forecast changes in demand in a more sophisticated way (the **aspirations** part of the supply function). Also, gains could be had from speeding up the actual building process itself. Suggested improvements in this context could be for examples:

1 Having 'off the shelf' plans.
2 Quicker methods of construction (the **technology** variable in our supply equation). For example, more prefabrication could be used (one would have to consider whether this could be done in aesthetically pleasing ways), and improvements in site labour productivity could be made, (see the discussion on productivity in the construction industry in Chapter 8).
3 Having stores of available building land (land banks).

The list could go on. However, it must be remembered that such increased preparedness can be risky, and certain aspects of it will either cost money or represent a significant opportunity cost in terms of 'tied up' capital. For example interest payments on borrowed monies to pay for a speculative land purchase could become very significant if the land is not quickly used say due to an unanticipated downturn in the market.

Linking these ideas in with demand, we will see why housing markets can be relatively volatile at both the local and national level due to demand fluctuations: as can be seen from Fig. 3.1, any changes in demand will cause large fluctuations in price. To see why demand changes over time we will initially look at our basic demand formula before going on to develop some further specific variables applicable to this market.

1 Firstly, we would expect our normal **price** relationship to hold in the case of housing, giving us a downwards sloping market demand curve. Simply, at any one time, with given demand conditions, less people will effectively be able to demand houses when prices are high, and more people will wish to purchase when prices are low. Moreover, at low prices some people may be encouraged to purchase more than one dwelling. Other houses purchased could be used as holiday homes or may be placed on the rental market as an investment.
2 Another important variable to consider is obviously **real incomes**. Essentially, the higher real disposable incomes are the more money

people have available for the purchase of owner occupied housing. This is especially true in countries where your ability to obtain a mortgage to help you to purchase the property is income dependent. Normally the total mortgage finance available to you for house purchase will be based upon a certain multiple (say three times) of your annual gross earnings. Thus, for example, if incomes were to increase significantly in an area, say due to the prosperity of local industry or business, it is likely that more people could afford their own home, or indeed purchase other houses for investment purposes. Coupled with this is the fact that people on high incomes are likely to view rented accommodation as an 'inferior good', and thus switch into owner occupation as soon as they can afford to do so. To fully appreciate the impact of income as a variable one should note that despite short run fluctuations, the **house prices:earnings ratio** in most countries is relatively static over time. For example, a house prices:earnings ratio of 3.5 signifies that house prices in the country in question are normally three and a half times that of the average income. Obviously if one was examining this issue in less developed economies a more disaggregated analysis would have to be conducted to reflect the huge divergencies in incomes and standards of living that often occur in such countries which typically exhibit 'dual economies'. Generally, however, increases in incomes will lead to increases in demand for owner occupation of residential property.

3 The substitutes available to you are also an important consideration. That is, if there is little in the way of public sector or private sector provision in the rented sector you may be forced into the situation of buying your own house. (This point is especially true in countries where the 'extended family system' does not operate and people do not tend to live with their parents.) Importantly, when we look at the issue of substitutes we should not just examine the number of available rented houses or flats, but also look at the relative prices of such accommodation, their quality, their location and so on, as these will all be key considerations when determining their real value as alternatives to owner occupation. The availability of substitutes to buying one's own home will obviously take away a lot of demand from the owner occupied market as people rent accommodation rather than buy it.

4 Taste is an important variable also. In countries where home ownership is felt to be an important social goal, owner occupation is likely to be high irrespective of price, thus we tend to get demand inelasticity. In fact the desire for home ownership will put upwards pressure on house prices as people compete with one another to acquire housing. Such 'taste orientated' demand can be enhanced further if there is a campaign, for example induced by government, to create even greater levels of home ownership.

Obviously this initial set of variables is not an exhaustive one, and other factors now need to be examined such as demographic change, and government policy at both the national and local level.

5 Under the heading of demography we can consider a variety of wide ranging issues:
(i) Housing demand is likely to increase dramatically about twenty to twenty-five years after a country has experienced a large increase in its birth rate. That is, demand increases will occur after a country has experienced a 'baby boom'. This is because, those born in the baby boom are expected to be in a situation to become 'first time buyers' in their early twenties as they gain employment and start to have families of their own. Specifically this tends to cause heavy demand for small 'first time buyer' dwellings thus putting upwards pressure on the general level of house prices as existing occupiers take the opportunity to 'move up the ladder' to larger homes.
(ii) A second common demographic factor in many countries is that they have experienced alarming increases in the divorce rate. This leads to more and more single people requiring single accommodation, whereas previously a couple would have lived in the same house. Linked to this is the modern trend to marry far later on in life. This trend has increased the demand for smaller housing units.
(iii) Another observable phenomenon in many nations is the rise in the number of elderly people, in particular elderly people living alone. This is primarily due to better living standards, improved medical technology, and the emergence of the 'nuclear family' system. This results in many more houses being occupied by the elderly, and thus this part of the stock is not being released on to the market.
(iv) Finally, some areas, or regions, may experience changes in demand due to net immigration or emigration. For example, areas that are very prosperous, say due to newly found economic prosperity, may experience increases in demand as more people move to the area to take advantage of its economic climate. Thus, huge increases in house prices will occur due to the likelihood of corresponding supply inelasticity. Conversely, of course, house prices may slump in areas where unemployment is high and people move away.

6 Government policy is also an important variable to consider. For example, demand may be encouraged by general mortgage interest tax relief, or even local authority renovation grants. But, just as importantly, although not necessarily as direct, is general government fiscal policy and monetary policy. Fiscal policy is where governments attempt to manipulate the economy by changing the level of government expenditure and/or taxation. Monetary policy is where governments attempt to manipulate the economy by changing the cost and/or availability of credit. (See Part 3 on macroeconomics and government policy.) For example, changes in the rate of interest will have an impact upon the cost of a mortgage. Thus, if an individual feels that they can afford to purchase a house with a 95 per cent mortgage at an interest rate of 10.5 per cent, they may be prevented from doing so if interest rates progressively rose to say 15 per cent, as this would greatly increase the level of

monthly repayments. Therefore, rises in the interest rate or, indeed, expectations of increase in the interest rate in the near future, may well curtail demand.

After considering the general workings of this market it will be useful now to briefly consider the effectiveness of it as a system with respect to the allocation of housing. Such a debate is felt to be necessary as we should never automatically assume that the market is the only, or best solution in this case: a substantial degree of public provision does exist in some countries.

Commonly cited arguments in favour of the market system are:

1 It should enable people to choose a house according to their own individual preferences, subject to the constraint of their income and their ability to raise mortgage finance.
2 No expensive government machinery is required to build dwellings, allocate them, and manage them.

But, in reality the 'achievements' of the housing market are potentially doubtful in as much that it is common to observe people unwillingly living in shared dwellings, or people living in accommodation that lacks basic amenities and/or is in serious need of repair. In fact, at the extreme, 'homelessness' is still a problem even in the most advanced of economies. Thus, it would appear that a degree of **market failure** occurs. That is, the market fails to fully operate in the manner that we expect. Potential areas of such market failure in the housing market are touched upon below:

1 Supply inelasticity (as discussed above pp. 46–47). Essentially supply inelasticity creates problems of rapidly increasing prices with little change in the existing stock to accommodate rising demand. This can reduce labour mobility, for example, as people cannot afford to move from one area to another as house prices increase in advance of them. Moreover, it makes prices volatile so that owners cannot be too complacent about the future value of their home.

2 As housing is an expensive commodity few can purchase a dwelling outright and therefore need to borrow finance. To cater for this, lending institutions, such as building societies, have come about in conjunction with similar facilities being offered by the banking sector. This, in itself, is not a problem, but such institutions tend to be understandably cautious and unwilling to lend to what they perceive as high-risk cases. Throwing 'caution to the wind' would obviously undermine the confidence of shareholders and depositors in the management of such institutions. However, many argue that excessive caution exists, and that this can lead to discrimination. For example, it has been known in the past for building societies to practice 'red lining'. This was a policy of demarcating parts of a town or a city as areas of 'high risk', and thus refusing to lend against houses in such areas. In fact, some have claimed

that such a policy of 'red lining' was not merely geographical but also discriminated against the ethnic origin of the loan applicant.

3 Another problem to consider is that of **imperfect information**. That is, with most goods that we buy, we buy them so often that we have a pretty good idea of their ruling market price. Moreover, most retail outlets are in heavy competition with one another and therefore prices are maintained at roughly the same level throughout the market. If we did not have such (perfect) information some sellers could charge excessive prices and accumulate abnormal profits. (See Chapter 7, section A concerning Perfect Competition.) Such imperfect information is widespread in the housing market, essentially because most people only occasionally purchase a house. Moreover, the costs of obtaining such information are considerable, for example time taken off from work for purposes of 'house hunting'. New immigrants to an area are likely to suffer from this problem to a greater extent than local people as the former are, by definition, less aware of the local property market. It may be argued that estate agents can help to reduce this type of market failure by being the source of accurate market information. However, some argue that it is in the interests of the estate agents themselves to try to keep prices artificially high. This allegation is made on the grounds that as estate agents usually receive a percentage of the house price on completion of sale, the higher the sale price the higher will be their income. If, for example, there are only a few estate agencies in a town they may attempt to interfere in the market to their joint benefit by secretly colluding together to 'decide upon' prices for the range of different types of property that are for sale. However, such behaviour, if it does occur, could 'backfire' on the perpetrators as housing demand declines in response to higher house prices.

4 The market may not fully take into account any associated **externalities**. An externality is created when the behaviour of one person, or group of people, can affect the welfare of another person, or group, other than through the price system. Externalities can be **positive** or **negative**. In the former instance benefits will be derived by some due to the action of others, whereas in the latter case the actions of some will affect others in a detrimental way, (see Chapter 15). For example, if a householder lets their property deteriorate, due to a lack of repair and maintenance, it may become an 'eyesore' and reduce the utility derived by the neighbours from their own housing. Another negative externality could be caused by a neighbour continually hosting late night, noisy parties. Such externalities may not be taken into account by the market mechanism. In the case of a noisy neighbour, for example, a potential purchaser may be unaware of this problem if viewing the property during the daytime. However, although obviously not problematic, there are also potential positive externalities. For example, on purchasing a property you may be unaware that your neighbours are planning to extensively landscape their front garden and improve the exterior of their house. Such activity

could not only provide you with a more pleasant outlook, but it could actually enhance the value of your own house as the street looks more pleasant and attracts other purchasers.

5 We should also recognize that housing is obviously fixed in location and thus cannot be moved from one place to another. Therefore, unlike many goods, shortages may persist in some areas whereas surpluses could occur in others. It is technically possible to take a house down and rebuild it elsewhere, yet this is an expensive exercise that is rarely performed.

6 The market may fail to adequately cater for those who are on very low incomes and cannot afford to enter the market even on the 'bottom rung' of the housing ladder. Because of this problem it is often said that the market is capable of achieving an *efficient solution* but not an *equitable* one.

In conclusion it must be said that although there are problems with the market system we should not just reject it as a suitable means of allocation. Rather we could:

1 Link it with some form of government control or provision; and/or

2 Find ways of reducing or even eliminating the sources of market failure discussed above.

It should be noted that these two concluding points should not be seen as being mutually exclusive.

3B The market for land: (with specific reference to building land)

Many textbooks will treat land as a 'special case'. They will often state that as there is only a given amount of land in any country its supply is completely fixed, and will thus be depicted as a perfectly inelastic supply curve. Therefore, according to this view, any change in demand will not lead to more land being supplied as no more land is available. Thus, the only effect would be radical changes in land prices. This is contrasted with the situation of most goods and services whereby increases in demand will encourage producers to produce more, and thus supply more, of the good or service in question. However, although it is correct to say that the *total* amount of land in a country is fixed (apart from small changes due to erosion, or land reclamation for example), this is merely the stock of land and does not reflect the amount traded or the incentive to trade, that is the flow of land traded in the market at any given period of time.

Therefore, if we are looking at building land in particular, we should not normally expect this to be completely inelastic in supply. Rather, we should expect a normal, upwards sloping supply curve as seen in Fig 3.2. The supply curve in Fig. 3.2 implies that there may be people who own suitable land, which the planners have categorized as land available for

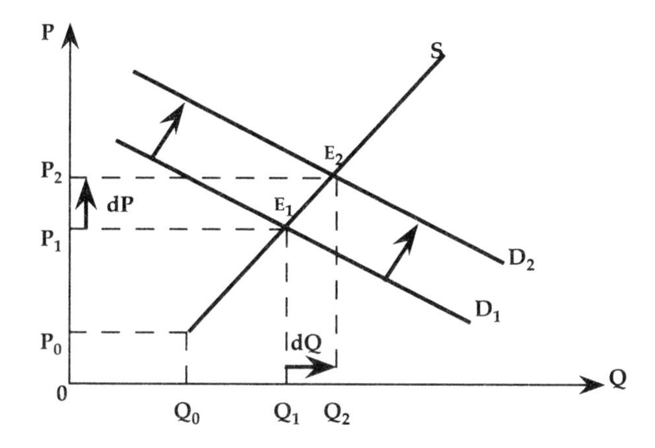

Figure 3.2. *The supply of building land and changes in demand*

development, and who are willing to sell to construction firms when the price is anything in excess of P_0. For example, if the level of demand causes land prices to reach P_1, quantity Q_1 will be traded. However, there may be many landowners who will only sell their land if offered an even higher price. For example, some may well wait until the demand for land has increased to D_2 giving us a new, higher, equilibrium price of P_2. This higher price would encourage Q_1Q_2 more land to be sold on the market. Let us imagine, for example, that you live on a one hectare plot of land, and that you are allowed to subdivide the existing property into three further building plots (see Fig. 3.3). Whether you do this or not will greatly depend upon the price that you are offered for these building plots (assuming that you are not going to develop the land yourself). Referring back to Fig. 3.2, if you were offered a low price such as P_0 you would probably retain your property in its entirety valuing the space and seclusion that the land in its existing format provided you. However, if land prices were to be bid upwards, say to P_1, due perhaps to general economic prosperity, or an increase in demand for housing in that area, you may be tempted to sell off one of the pieces of land that will not affect you too much with respect to loss of view etc., providing that adequate access and services could be provided. Such a plot is likely to be either plot A or plot C depending upon your specific preferences as they are less in the line of view of the existing house. However, if the demand for building land increased yet further, giving a new equilibrium price of P_2, say due to a housing boom, you may be encouraged to sell off all of your available land. Indeed, it may be worth selling all the land and redeveloping the original plot in a more extensive manner. Thus, as with the productive process or with the provision of services, price will give you the incentive to sell. That is, the higher the price, the more you are willing to sell. This is the first law of supply (see back to Chapter 1). Likewise, if we were considering agricultural land redesignated as potential development land, the farmer, if given the choice, is likely to sell off marginal land first before he even considers letting go prime arable

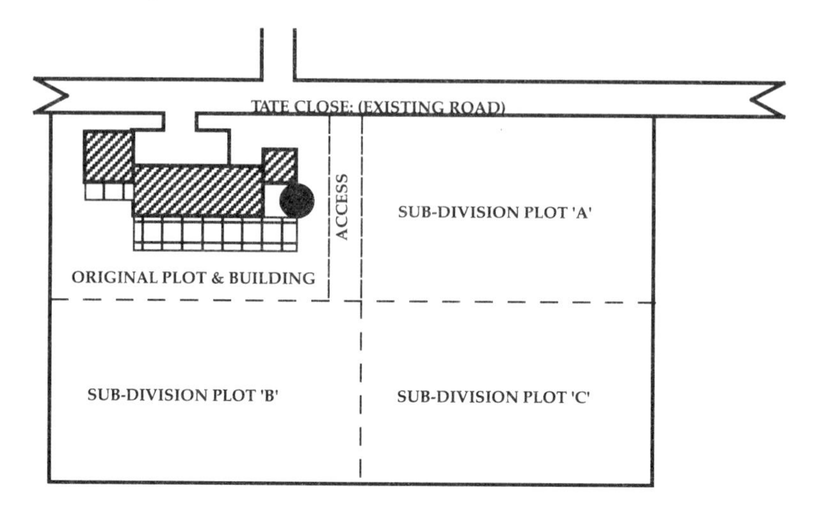

Figure 3.3. *The supply of land due to changes in price – a case study*

or grazing land. Moreover, the illogicality of the perfectly inelastic supply curve (as shown in Fig. 3.4) is that it implies that you would be just as willing to sell your land at P_1, as at P_2, as the same quantity (Q^*) is traded at both these prices.

Obviously though, the supply of land in many circumstances, especially at the local level, is likely to be highly inelastic inasmuch as the number of landowners willing to sell (assuming no compulsory purchase orders have been served), general planning constraints, and tight 'green belt' policy will limit the amount of land available for development at any given time. However, normally, even in a highly developed urban environment, some land will flow on to the market as sites are used more intensively, old buildings are demolished, due for example to functional obsolescence, and

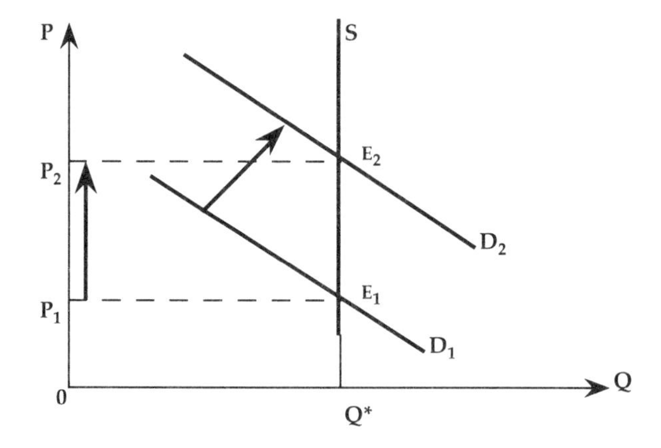

Figure 3.4. *The illogical nature of the perfectly inelastic supply curve*

there is a general expansion of urban areas. Moreover, as suggested above, landowners are likely to behave like any other entrepreneur and would not necessarily sell simply because they had a vacant plot of development land. Why sell land in the depths of a severe economic recession when such land could become prime commercial or industrial land commanding a high price during an eventual upturn in the economy?

Another important concept to grasp is that the demand for building land is a **derived demand**. That is, construction firms, unless they are extremely eccentric, do not just demand land for the sake of it. They demand land because they can build upon it now, or at some future date, and hopefully sell the completed development at an acceptable profit. Thus, the price that they are willing to offer the landowner will depend upon the state of the market for completed buildings, or the future expected state of the construction market in the case of the speculative builder, or for a development that is not pre-sold and will take some time to complete. For example, take the case of house builders. During a boom in the housing market when house prices increase rapidly, the house builder will realize that as demand is high, they could build and sell more houses than normal, and at higher prices, which in turn should enhance the firms profitability. (See Chapter 5, concerning costs and revenues.) Thus, if the building firms feel that the boom will be sufficiently sustained, they will look around for more available building land. Consequently, the demand for building land increases, and land prices rise. Therefore, the increase in land prices has originated from the original rise in house prices. This is diagrammatically shown in Fig. 3.5. Here, as house prices rise from P_1 to P_2, due to a rise in housing demand from D_1 to D_2, quantity Q_1Q_2 more houses are built upon Q_1Q_2 hectares of land. To tempt landowners to part with Q_1Q_2 hectares of land, land prices increase from P_1 to P_2. The price that a builder will be willing to offer the landowner is frequently referred to as the **residual**. That is, what is left over after profits and all construction costs have been taken

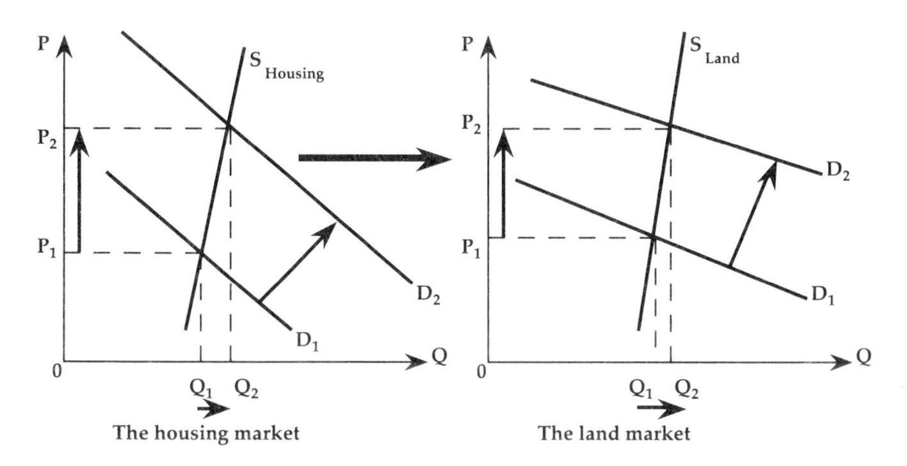

Figure 3.5. *The derived demand for land*

into account. Thus, if house prices are higher, for example, higher profits may be taken, and indeed there may well be upwards pressure on building costs as more labour and building materials are demanded, but the amount of money left over to bid for land is also likely to be greater. In other words, with reference to Fig. 3.6, and if we continue the example of escalating house prices, we see (by viewing the graphs from left to right) that the demand for housing has risen and the number of new houses being built has increased from Q_1 to Q_2. Therefore, even if the builder's profit margin has increased, and costs have risen, it is likely that more will be left over for bidding for building land. That is, the excess left over is now greater. Diagrammatically this excess or 'residual' can be shown by area $X_2P_2E_2D$ being greater than area $X_1P_1E_1A$, where the areas OC_1BQ_1 and OC_2EQ_2 represent builder's costs, and areas C_1X_1AB and C_2X_2DE represent the builder's profits in each instance. Thus, quadrangles $X_1P_1E_1A$ and $X_2P_2E_2D$ will be the areas under the demand curve for land in each instance as they simply denote the amount of money at the builder's disposal to bid for building land. This can be seen in Fig. 3.7 where the residual (shaded) areas from Fig. 3.6 have been transferred on to Fig. 3.7. Note that the sizes of the transferred areas have been exaggerated for the purpose of clarity.

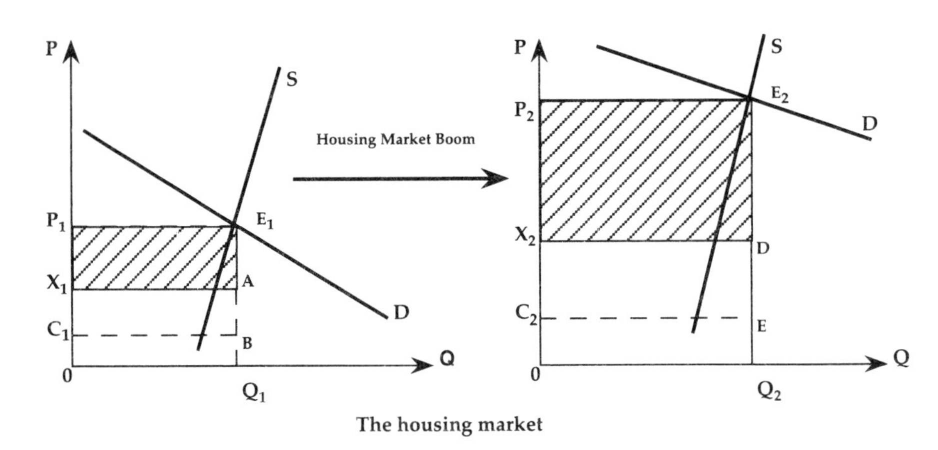

Figure 3.6. *The notion of the residual*

Variations to this theory can be examined, such as the fact that the land used for a current development may be acquired from a previously obtained land bank. In such a case, the residual simply represents the amount of money available to buy the next area of land for the next development.

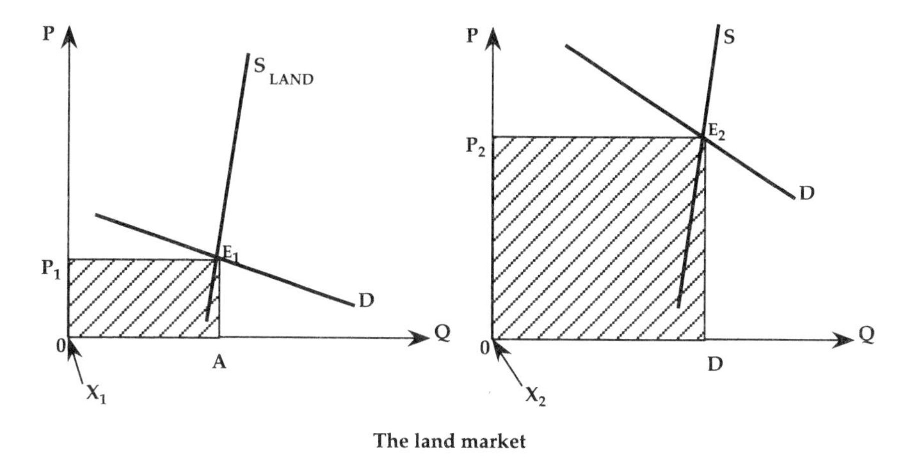

Figure 3.7. *The demand for building land*

3C The market for construction components

The decision by the builder to use certain building materials will depend upon a wide variety of variables such as:

1 The range of materials available on the market to chose from. The variety of materials available will largely depend upon such factors as the number of firms in the materials supply sector, the degree of competition between these firms, and the level of import penetration of such goods.

2 Building Standards. Obviously the materials used must conform to existing policies on structural soundness, health and safety. Materials that are known to be structurally unsound or hazardous to health must not be used, and therefore act as a constraint in the decision process.

3 The quality of materials. A great range of quality is available for the majority of construction components. A builder could select lower quality materials (as long as they conformed to existing standards) in an effort to save on costs. Remember, however, that many components such as wall ties, plumbing and so on cannot be seen by the client, and therefore some aspects of quality may only be noticed by their visibility to the buildings users. Alternatively, the builder could select high quality, high cost, fittings such as bathroom suites, and kitchen units (in the case of housing) in the hope of advertising these features as a selling point of the house. The quality of materials could also depend upon the type of building under construction, whether it be industrial or residential, and indeed the estimated economic life of the building would be an important consideration. On the latter point it would be unwise to fit out a building at a high cost if it was expected that it would be functionally obsolete within ten years, for example, and then demolished to make way for a new building.

4 The cheapness of materials. Obviously the cost of materials is often married to their quality and therefore the discussion directly above is relevant. However, cost and quality are not always the same. For example, during a recession, because of falling demand, many construction components may well drop in price although their quality is the same. Furthermore, relative costs may change as new production techniques are found, and new producers arrive on the market to compete with existing suppliers. An example of the latter case is cheap imports from abroad that 'under cut' existing domestic suppliers.

5 Consumer Demand. If we are dealing with visual materials builders would be wise to ensure that they keep in line with current consumer tastes. In fact, installing popular items can, of course, be used as a selling point of the building in question. Alternatively the builder may allow the client to select a variety of fittings according to their own specific needs and tastes. For example, many house builders allow the purchaser to select such items as tiles, bathroom suites, and kitchen units.

All of these variables, and others, will affect the demand and supply of any material that we select for the purpose of an explanatory example. It would be a laborious exercise to go through each of the above points in turn in any great depth but they are worthy of consideration. In the following example I have chosen two products that could be seen as near substitutes as they perform similar tasks: namely metal window frames and wooden window frames.

In a building 'boom' more buildings would be constructed and thus more window frames required. Moreover, if the building boom was (as is likely to be the case) due to an accompanying economic recovery, more people could also afford new replacement windows on existing buildings. Window replacement may be seen as desirable if the existing windows are in a poor state of repair, or if the new windows have superior qualities such as enhanced energy efficiency, or better draft and noise exclusion. Thus, the demand for both types of window frame is likely to increase, and depending upon their elasticity of supply, this will tend to have an inflationary effect upon the price of them. For example, if there were only a few firms in any area producing wooden window frames, they may be unable to cope, in the short run, with any dramatic increase in demand. As such, the only way that they could 'ration out' supply is via an increase in price as depicted in Fig. 3.8, where prices rise from P_1 to P_2 with only a very limited increase in production from Q_1 to Q_2. However, if imports could be found from other regions or countries, the supply curve would become less inelastic and therefore demand would be spread over more firms and there would be less of an impact upon price (see Fig. 3.9). Price competition is likely to exist between these firms anyway. Alternatively though, an increase in demand could be diverted into a substitute good such as plastic window frames or metal window frames, if insufficient new supplies of the original good could not be found. Another possibility is that one product, for example metal window frames, could become unfashion-

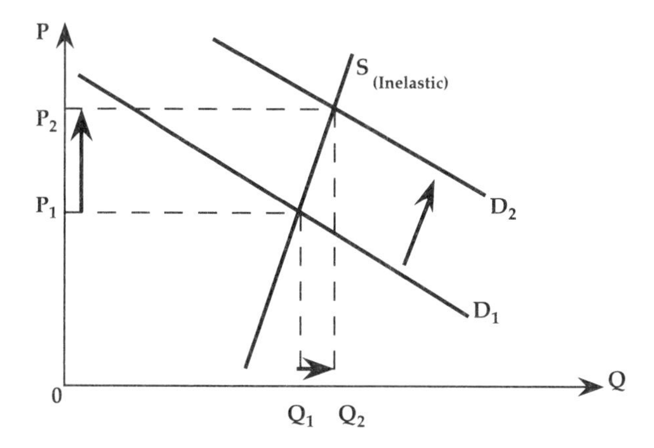

Figure 3.8. *Increasing demand for wooden window frames – the case of limited supply*

able in the public eye. This is likely to create a drop in demand, and thus price, for metal window frames as depicted in Fig. 3.10, and an increase in the demand, and hence price, of wooden window frames. Obviously such a situation would depend upon many factors – the supply of each product, the degree of substitutability between the goods, and the potential existence of near substitutes such as plastic window frames. Moreover, the limit to such substitution depends upon whether the consumer is willing to pay the cost of having the alternative material in the price of the completed dwelling. If they are not, one could reach a stage where wooden window frames became too expensive, (see Fig. 3.8), and either metal window frames would be installed, or cheaper, lower quality wooden window frames would be sought.

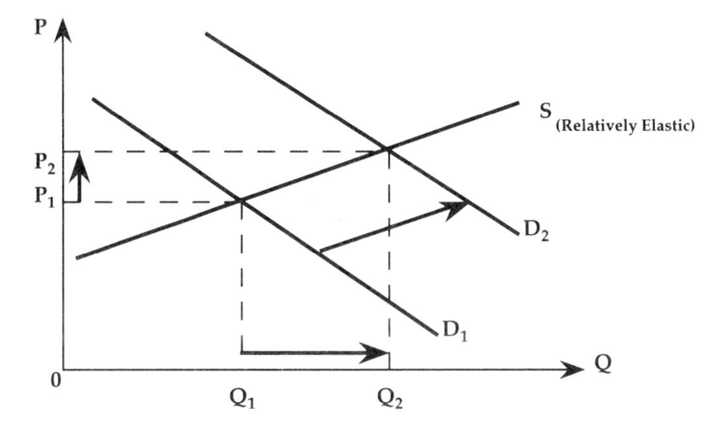

Figure 3.9. *Increasing demand for wooden window frames – extensive supply*

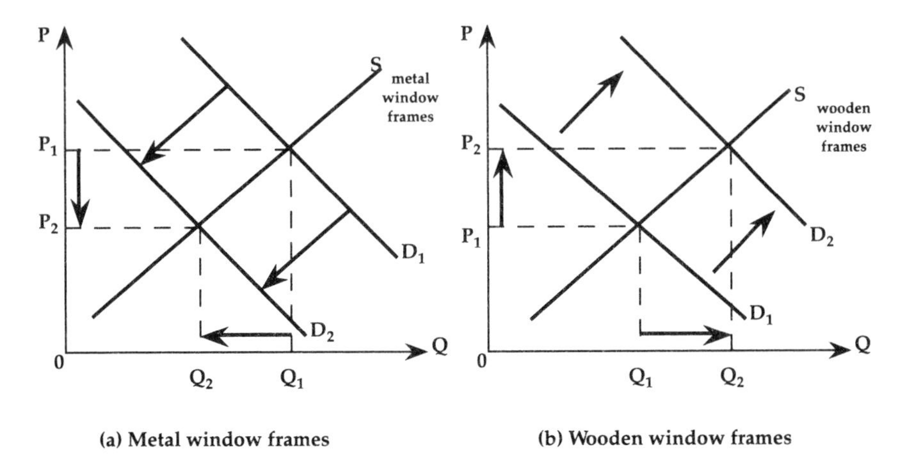

Figure 3.10. *Decreasing demand for the original good, leading to an increase in demand for a substitute good*

3D The market for building labour

Although it is true to say that the general behavioural aspects of labour will be similar whether you are a surveyor, carpenter, or a bricklayer, it would be naïve to treat labour under an all encompassing heading to discuss the behaviour of all labour within the economy. This is because different circumstances will affect different trades and professions over time, and as such, it is best to disaggregate them along these lines before initiating any comprehensive or meaningful study or analysis. Therefore, in this limited section I will concentrate on an example of one trade, that of bricklayers. Here I will illustrate how a variety of factors may affect their employment, and in turn one may see the potential impact of these conditions upon other related issues such as productivity, the design of the building, the materials used, etc.

As a starting point, let us imagine the overall market for bricklayers. Just as with any other market, there will be a demand for bricklayers, (a derived demand in this instance). In many countries houses are built in a 'traditional' manner with the outer walls constructed of brick. As such we would expect the demand for bricklayers to be highly inelastic as shown in Fig. 3.11 by the short run demand curve $DL_1(SR)$. Such inelasticity is likely because even if wage rates, or the amount paid to a bricklaying team per house built, were to be increased dramatically – say by effective trade union action or government imposed minimum wage legislation – building firms, at least in the short run, will have to employ roughly the same number of bricklayers to construct houses in the normal manner. Obviously, in the long run, the demand curve could become more elastic as bricklayers,

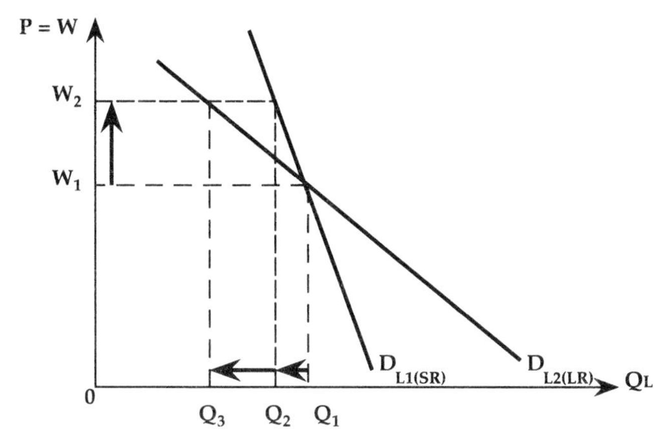

Figure 3.11. *The demand for bricklayers*

or bricklaying teams, could be laid off as building firms could react to the problem of rising wages in a variety of ways. For example,

1 Firms could insist upon higher productivity rates from existing bricklayers or bricklaying teams.

2 Firms could assist such productivity enhancements by ensuring that any potential supply constraints which could prevent the bricklayers working as quickly as required were reduced or eliminated. Improvements could be along the lines of:
(i) Reducing time delays in getting materials to the workers.
(ii) Reducing site wastage so as to ensure that adequate materials are available when required.
(iii) Ensuring that production is organized efficiently so that the time a bricklayer needs to spend on a house is kept to a minimum. In other words, the site manager should try to arrange the sequence of construction in such a way that a bricklayer does not have to return to a house to re-do part of the job that has been damaged or altered by another subcontractor for example.

3 Firms could simplify the design of their houses so that less bricklaying, or at least less complex, time consuming, bricklaying is required. For example, at the extreme, a simple 'box shaped' house will use less bricks than a house of a more intricate design and layout. However, firms would have to be aware that such cost cutting actions could also reduce the price of the house and thus revenue received, as there may be less demand for such 'utilitarian' housing units.

4 Changing, or adapting, the method of construction so that less on site bricklaying is required. This could be achieved, for example, by the use of more prefabricated materials and units, and/or using alternative materials – more glazing for instance.

Referring again to Fig. 3.11 it can be seen that in the short run if, for example, trade unions pushed the wage rates up from w_1 to w_2, firms could not simply lay off workers as governed by the demand curve $DL_1(SR)$. However, as discussed above, in the long run firms could alter their methods of production, and thus the demand curve would become more elastic. Therefore, with the initial demand curve wage increases would cause the laying off of only Q_1Q_2 bricklayers, whereas, in the long run, as demand moves to $DL_2(LR)$, a further Q_2Q_3 bricklayers could also be made redundant.

On the supply side (see Fig. 3.12) the supply curve will represent how many people at any given time, wish to be, or are qualified to be, bricklayers. Some of these people will be willing to work at a relatively low wage rate such as w_1, whereas others will only be able to supply their labour at relatively high wage levels such as w_2. The diagram suggests that nobody would work for a wage less than w_0, as such a wage would be deemed too low to attract people into this particular trade. Presumably at a wage that low people would be better off seeking alternative employment, or receiving state unemployment benefits (if applicable).

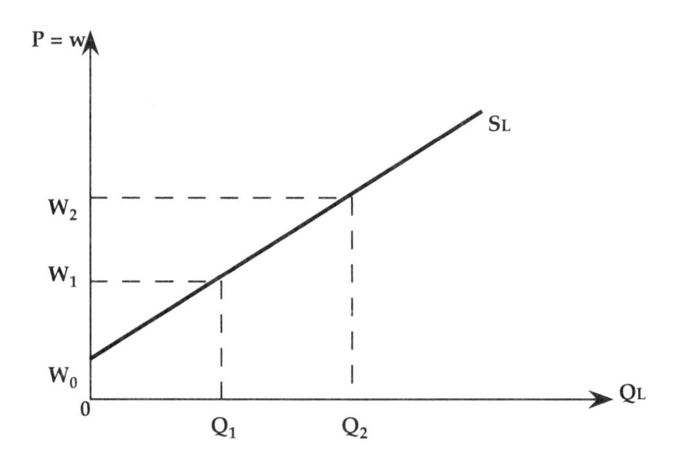

Figure 3.12. *The supply of bricklayers*

Bringing the two concepts of demand and supply together we will arrive at the market for bricklayers as shown in Fig. 3.13. This situation would give us a market wage of w_1 and Q_1 bricklayers would be employed. As with all markets, this outcome is likely to vary on a local scale depending upon local conditions of supply and demand. For example, there may be a shortage of bricklayers in an area where there are many building sites under construction. Such a situation is likely to drive wage levels to a higher level as demand is high and supply is constrained. In Fig. 3.13 the area $0w_0EQ_1$ represents the **transfer earnings** of bricklayers, and the area w_0w_1E, their **economic rent**. Transfer earnings are the payments necessary to keep a particular factor of production in its present use. For example, if

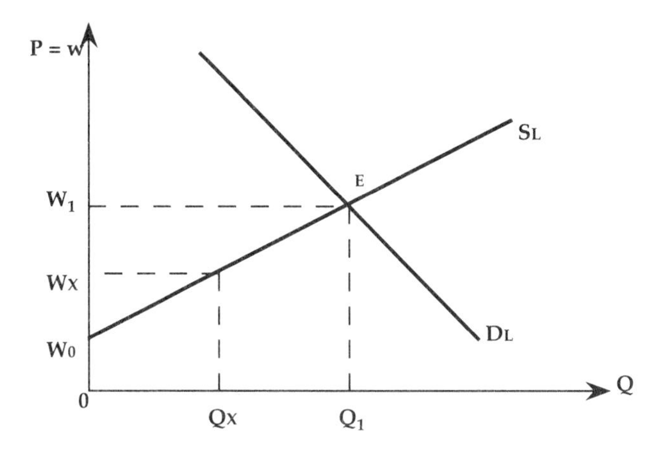

Figure 3.13. *The market for bricklayers*

the bricklayers represented by Qx did not receive a wage of at least wx they would elect to work (transfer) elsewhere, or become voluntarily redundant as presumably they would be better off working in another trade or living off state benefits (if available). Thus, transfer earnings represent the **opportunity cost** of working in this particular trade as the worker could earn this amount by working elsewhere. Economic rent, on the other hand, is an added bonus, a payment over and above that necessary to keep the worker on (as a bricklayer in this case). Economic rent only comes about because of the fact that the high level of demand has attracted people with high transfer earnings, such as those workers represented by Q_1 on the diagram. Note that worker Q_1 has no economic rent whereas workers (who may be less versatile on the job market) such as Qx have a significant economic rent. The worker represented by Q_1 may also be skilled as a plumber for example, and as such they will only be willing to be a bricklayer if the returns they receive from this are greater than those received from plumbing.

It must again be stressed that this model is a dynamic concept and not a static one. That is, both demand and supply could, and will, change over time. For example, imagine the house building industry going into recession, say due to a wider economic recession caused by high interest rates. In such a situation we would expect building firms to react to depressed demand by building less houses, and therefore less bricklayers will be demanded. Thus, wages are likely to drop (say from w_1 to w_3 on Fig. 3.14) as the demand for bricklayers drops from D_1 to D_2, and a new market equilibrium is achieved at a lower level with Q_1Q^* bricklayers being laid off. It is important to note, however, that quantifying such a drop in demand may be quite difficult as firms could react to a recession by changing the 'mix' of their developments and simply build different types of houses. That is, they could find out which houses are least affected by the decline in people's purchasing power: 'executive style', large houses, may be harder hit than the 'first time buyer' market, and therefore builders will seek planning permission to build more smaller units. Thus, the total

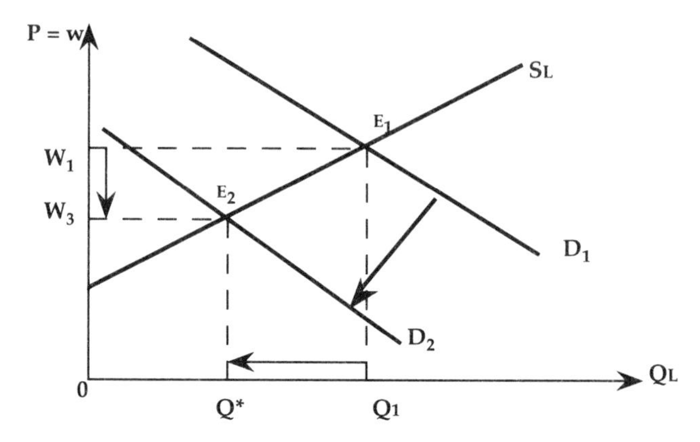

Figure 3.14. *The market for bricklayers: a building recession*

impact on the employment market may initially be quite small. Moreover, another consideration is that developers could try to protect their situation in such a depressed market by selling houses of a cheaper form of construction and finish so as to retain some profit as house prices decline. Conversely, however, in such situations some firms have 'gambled' by building houses of higher specification and cost so as to find a particular niche in the market. Both of these reactions to a recession may help to protect the prospects of the bricklayer. Obviously, the above analysis could be reversed if we were looking at a housing boom. Such situations lead to rapidly increasing demands for site labour. If such labour is in relatively short (inelastic) supply this can lead to a rapid escalation in site labourers' wages. Imagine the simplified situation where we have a town that was experiencing a housing boom, yet there were only a given number of bricklaying teams in that town (say Q^* in Fig. 3.15). This could be represented by a completely inelastic supply curve for bricklayers in the very short run

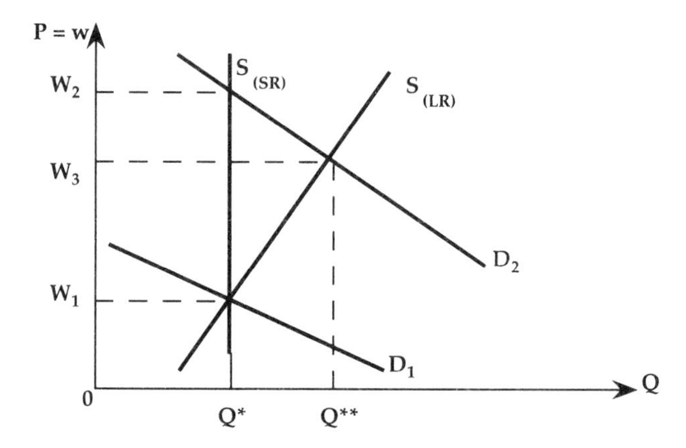

Figure 3.15. *The demand for bricklayers: a building recovery and boom*

as shown by $S_{(SR)}$. Therefore, as demand increased from D_1 to D_2 existing teams could work harder and earn overtime payments equal to w_1w_2. Such high wage rates, or returns, are likely to attract others to become bricklayers in the area, whether they become trained in the trade, or they travel in to work from other areas. If this did arise, the supply of bricklayers, in the long run, would be increased so as to reflect the increased levels of competition from other bricklayers entering the market giving a new supply curve of $S_{(LR)}$. Building firms could now choose from a greater pool of labour and wages would be driven down to w_3. The end result would then be that less overtime would be done, and Q^*Q^{**} more people would be employed.

It would be a useful exercise for you to think of other changes that could occur in this market, such as improvements in productivity, and follow them through in a logical, diagrammatic form as suggested above.

— 4 —

Government intervention in land and construction markets

This chapter is designed to introduce you to various forms of government intervention that can be imposed upon the free operation of the market. It is useful at this stage to give you an *insight* into how construction related markets are, or can be, manipulated by the public sector. However, more detailed analysis appears in later stages of this book (see Parts 3 and 4) where we consider the role and impact of government macroeconomic policy, and we look into the problems of market failure in more detail.

Essentially, if the market were to operate without intervention we should, in the absence of any market failure, arrive at an efficient, free market, equilibrium. However, there are numerous cases where such a solution may be viewed as inequitable, or even unsafe. Thus, under such circumstances corrective government action is taken in an attempt to improve the end result. This public interference could be taken at the national or local level. For purposes of illustration at this stage I have selected three areas of government intervention to briefly analyse below:

1 Taxation and subsidization in the built environment;
2 Regulatory policy: Health and Safety in Buildings, and Land Use Control;
3 Price control policies in the built environment.

(A) Taxation and subsidization and the built environment

(1) Taxation

There are a wide range of taxes that can be imposed by both central and local government on both consumers and producers. The purposes of such taxation can range from the raising of revenue, to the regulation of production or activity for the good of society in general.

(a) Taxation upon the producer

The government may impose taxes, or additional taxes, upon a good or service in order to attempt to reduce the amount supplied on to the market. Such taxation imposed upon the producer will, *ceteris paribus,* effectively increase the producers input costs, (remembering that supply is partially a function of 'input costs'), and therefore decrease supply at any given price. Such a leftwards shift in supply will have the effect of increasing the price of the product and thereby reducing demand. This movement of the supply curve can be seen in Fig. 4.1. Here, if the market were allowed to operate freely, we would reach an equilibrium at E_0, with a price equal to P_0 and the quantity traded equal to Q_0 as shown by the interaction of the demand curve D and the supply curve S_0. However, a newly imposed, or increased, tax will raise the production costs of the supplier by the amount of the tax, and therefore shift the supply curve to the left. Thus, less will be produced at a higher price. Note that normally, before the imposition of the tax, the supplier would have been willing to supply Q_1 for a price of P_{-1}, but now, to cover the tax, he charges P_1 for the same output. Thus, $P_{-1}P_0$ of the tax is paid for by the producer, and the rest, P_0P_1, is passed on to the consumer via a higher selling price. The ability of the producer to pass the tax on to the consumer is largely dependent upon the elasticity of demand. Such action could be seen as the driving motive behind the following examples:

(i) Taxing potentially hazardous, or non 'environmentally friendly' building materials. This should have the effect of increasing their price and thus encouraging builders to use alternative, safer, and more acceptable materials.

(ii) Increasing taxation on energy for lighting and heating such as gas and electricity. This should increase the price of such services to the householder and commercial user and thus encourage them to be more economical with the world's limited resources, as well as being more environmentally aware. Not only will higher energy bills make users become more cautious about energy wastage, but such a tax is also likely to encourage users to install energy saving devices such as double glazing, and cavity wall insulation, for example. Essentially rising energy costs would shorten the payback period of such investments and thus enhance their attractiveness.

(iii) Taxing a manufacturer, say for example a cement producer, whose factory is polluting the atmosphere. By imposing a tax on the company's

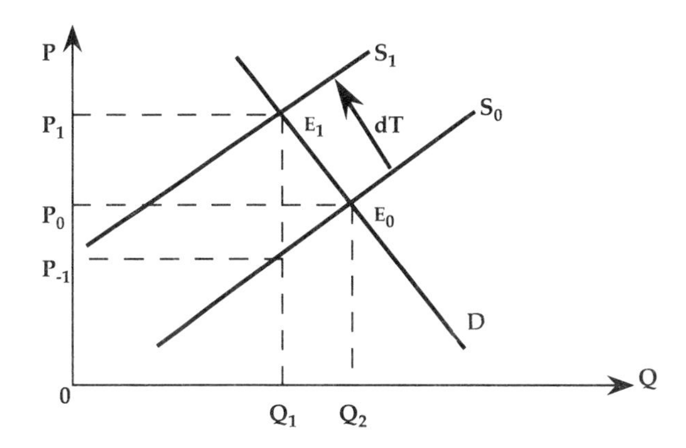

Figure 4.1. *Taxation on the producer*

product, the supply curve of the product would shift to the left, and thus less would be produced and demanded. (Consumers could perhaps buy from a supplier who is using a cleaner manufacturing process, and is therefore not attracting the tax.) In fact, such a tax, if high enough, could encourage existing producers to invest in less polluting methods of production so as to reduce the level of tax imposed. This sort of tax could therefore lead to a cleaner built environment.

(b) Taxation upon the consumer

Altering the level of tax that a consumer has to pay will obviously affect the level of income at the consumer's disposal. Therefore, the higher the tax, the less the disposable income of the consumer. As disposable income is a key element of demand this will decrease a consumer's ability to demand goods and services, and thus their demand curve will shift down to the left. This can be seen in Fig. 4.2 where demand has fallen from D_0 to D_1 effectively reducing output from Q_0 to Q_1. Therefore, by using an example from above, you could attempt to encourage householders to use less energy by reducing their spending power rather than increasing the price of the product. However, the effectiveness of such a policy would be largely dependent upon the consumer's relative elasticities of demand between different goods and services. Moreover, the situation is complicated yet further as the consumer could continue to use the same amount of energy as before, yet cut down on other essential expenditure such as that on food and clothing. Furthermore, the impact of such a policy could be delayed by consumers taking out loan finance in order to keep their consumption levels the same as before while they find it difficult to make economies, or cannot afford to install energy saving devices.

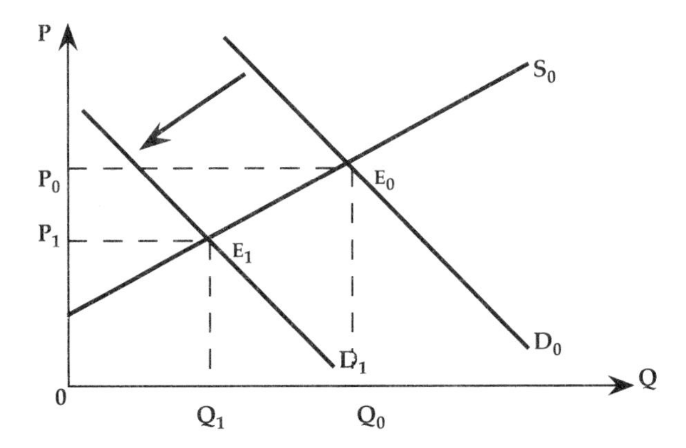

Figure 4.2. *A reduction in consumer demand due to higher taxes reducing disposable income*

(2) Subsidization

If the opposite effect of taxation was felt to be desirable by government (local or national) they could grant subsidies to the producer or consumer to encourage production and consumption of a particular activity.

(a) Subsidizing the producer

Monies aimed at increasing supply could be aimed at the producer who is providing a good or service which is beneficial to society. Examples could be:

(i) Manufacturers producing energy saving devices or materials could be given money to encourage higher output, lower costs and further research. As could be firms who are producing a whole range of environmentally friendly goods or services.

(ii) Manufacturers using methods of production which are now unacceptable in terms of pollution and/or safety, for example, could receive subsidies to pay for more modern equipment to replace the old, or to learn about new techniques.

(iii). Subsidies could be used to encourage firms to provide more, or more varied services if it were involved, say, in the field of public transport. For example non-profit making routes that are none the less essential for outlying communities, could be run with the help of such finance.

The impact of such subsidization can be seen in Fig. 4.3, where it is seen to lower the suppliers' costs of production, thus enabling them to supply more at any given price. Under these conditions the supply curve will shift to the right from S_0 to S_1, where more will be produced, and at a lower price, as output increases from Q_0 to Q_1, and price falls to P_1 from P_0.

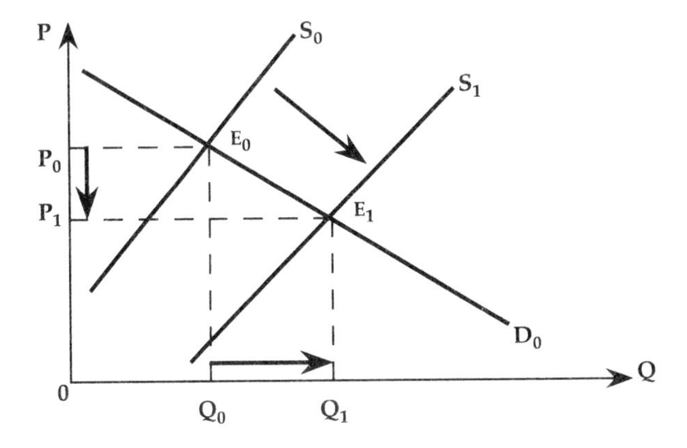

Figure 4.3. *The case of a producer subsidy*

(b) Subsidising the consumer

Here we could use the example of a local authority who, wishing to improve a depressed area, promoted an enhancement in the quality of existing privately owned housing stock in that area. To encourage people to improve their homes, home improvement grants, and/or renovation grants could be offered. Thus, people could obtain public money to increase the standard of their properties by undertaking improvements making it a nicer area for *all* to live in. Note that, as can be seen in Fig. 4.4, the detrimental side effect of this policy could be that items required for home improvements and renovation may go up in price reflecting increases in demand for them. The full effect of this price movement would depend upon the elasticity of supply.

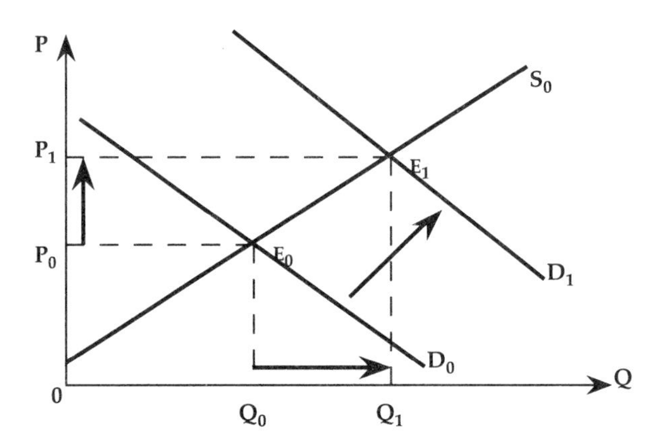

Figure 4.4. *Consumer subsidies giving rise to an increase in demand for home improvement and renovation*

(B) Regulatory policy: health and safety in buildings and land use control

Health and safety in buildings and land use control (planning), are two important areas that demonstrate where government *directly* intervenes in the market.

(1) Health and safety in buildings

Whether one is involved in manufacturing, construction, or the provision of services, research and development (R&D) leading to *technical innovation* is encouraged so as to:

1 Improve the product in order to make it more desirable for the consumer or end user. Such improvements should help to maintain, or even increase, demand for the product, giving rise to potential positive rev-

enue implications that should outweigh any increased costs of production, at least in the long run. For example, house builders may invest in techniques which improve the quality of their houses so as to attract more people to buy them.

2 Find a means of reducing the firm's costs so that it can become more competitive. For example the firm could find new, cheaper materials that should perform the same function as the original materials used. An instance of this would be a builder using a new, cheaper type of cavity wall insulation that had adequate properties so as to meet current energy standards.

3 Enhance productivity by finding faster, more efficient, methods of production or provision of service. This should enable the firm to meet the needs of the market more quickly and therefore recover its costs at a faster rate. Such a goal is especially important where large amounts of loan finance need to be repaid by the producer so as to avoid on-going interest repayments, as is typically the case with most construction projects.

Apart from the case of merely improving quality, all of the above actions would have the effect of shifting the supply curve to the right, as technology is a key variable in the supply function. Thus, technological change should enable more to be produced, provided, or built, at any given price. This can be seen in Fig. 4.5 where the supply curve has moved from S_1 to S_2 leading to an increase in the quantity traded from Q_1 to Q_2, and a reduction in price from P_1 to P_2. As can be seen from the diagram, such technological change could be passed on, in part, to the consumer, or end user, via lower prices, (or, possibly an improved product as well, although this is not necessarily the case if the technology has been designed to improve the producers' lot rather than the consumers'). However, the application of such technology to the productive or building process, for example, may be seen

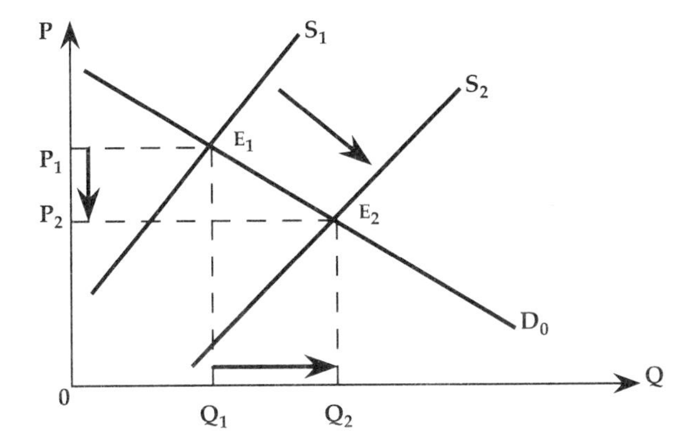

Figure 4.5. *Shifting supply due to technological innovation*

by *regulatory bodies* as being unsafe for the workers involved, or indeed they could be seen as unsafe for the consumer or end user. Thus, government *safety regulations, building standards,* or *manufacturing standards* could be imposed to either:

1 modify the new technology to make it safer and thus more acceptable; or
2 reverse the technological change completely so as to legally prevent it from being used.

In each of these instances the supply curve is likely to shift back to the left, *ceteris paribus*. This retreat of the supply curve would obviously be complete in the case of the new technology being found to be totally unacceptable as in (2) above. It is important to realize that many methods of production and construction can be used for a considerable amount of time before they are declared to be unsafe. Problems can be brought to the notice of regulatory bodies by the reporting or observation of frequent accidents on site, or structural failure in buildings for example. Therefore, such legislative reversals, as described above, may not be immediate and could thus lead to high levels of rectification work on existing buildings.

(2) Land use controls: land use planning

Land use planners may deliberately prevent the building, or at least limit the supply of certain types of building, so as to avoid the risk of commercial collapse, and at the same time to strive to achieve the highest possible level of welfare for society in any given area. For example, although a manufacturer may wish to locate in a particular area due to perceived advantages of location near a market, raw material, or major route ways, such development may be prevented, or restricted, if this were to produce a situation of *conflicting land uses*. For instance allowing heavy industrial development in a primarily residential area, or an area of 'outstanding natural beauty' would impose social costs upon society thus detracting from their welfare. Such social costs could be increased pollution, increased congestion due to heavy vehicle traffic to and from the factory, and the loss of green areas, etc. Planners may also reject such an application because they are aware of other similar future developments in the vicinity. Giving all the proposals the 'go ahead' may lead to an over supply of such businesses and buildings which could subsequently lead to business failure and dereliction. Not only may jobs be lost, but such resultant dereliction is likely to be viewed as unsightly and thus a cost to society, as well as a waste of resources. On the other hand, planners can positively encourage the supply of certain types of buildings so as to stimulate the growth of *complementary land uses*. For example, planners could try to create a business and/or industrial zone or 'park'. Here it is argued that, in such developments, not only can firms benefit from being near to their suppliers for cost reasons, but also by being near to their competitors they can enhance their awareness of the market. Such improved knowledge should facilitate discussions between firms that could lead to improved business prospects for all those concerned. These positive 'spin offs' are known as **agglomeration economies**. Note that land use control can occur at both the national and local level. On a national basis certain guidelines can be laid down so as to achieve a cohesive national socioeconomic policy, whereas local governments are concerned with examining individual cases as well as adhering to an overall central policy. For a more detailed examination of the economic rationale for land use planning see Chapter 15.

(C) Price control policies

For a variety of reasons central, or local government, may feel that the pure market result for some goods and services is an unjust one. As such, they may be able to seek the power to impose and enforce either:

1 a maximum price control;
2 a minimum price control.

Both of these policies will now be individually examined.

(1) Maximum price controls (price ceilings)

A maximum price control is the setting of a legal maximum price on a good or service whereby the providers of those goods and services are not allowed to charge above this imposed price. That is, a **price ceiling** is created and prices are not permitted to go through, or higher than, this artificial constraint. Such a ceiling would be imposed, for example, if the government believed that the market price of a particular good (say rented housing), or service (say fees for a valuation survey or structural survey needed to obtain necessary finance for house purchase), was inequitably high when looked at from the angle of the general consumer.

However, it is important to note that, as with most policies, they rarely achieve their goal without causing secondary effects which may be positive or negative. To demonstrate this we will now go through two applied examples specifically highlighting additional measures that need to be imposed in conjunction with such price interference in order to reduce or eliminate any potential ill effects of the original policy.

Example 1: The rented residential accommodation market

If the market were allowed to operate freely, we would get a series of equilibrium rents being achieved for various types of rented residential property. That is, the market for such property should be sub-divided into smaller sub-markets to reflect different types of accommodation and specific locations. For example, different people will be concerned with renting a flat in a town centre than those who wish to rent an 'executive' house in the suburbs. Thus, in order to concentrate the analysis on one sub market,

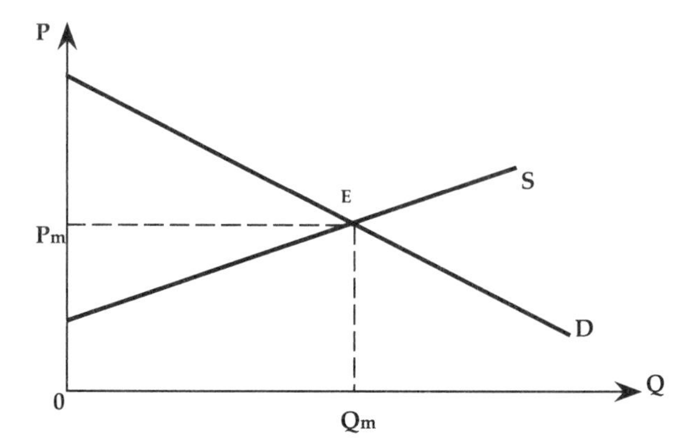

Figure 4.6. *The free market for city centre rented accommodation*

we will now narrow our area of study down to examining town centre properties, which in our example, will consist primarily of old, terraced houses. As we would expect, and as can be seen by from Fig. 4.6, there will be a demand for such properties and a potential supply of them as shown by the demand curve *D*, and the supply curve *S*. The demand curve, as usual, will be downward sloping since fewer will wish to pay rent for such accommodation if it is priced too high, yet such houses would attract many if the rent is sufficiently low. For example, a low rent could attract young people to move away from their parents' home as they could now afford the independence of their own house to rent. Likewise, low rents are likely to attract people on relatively low incomes, such as students, who could now afford to live with fewer people in each dwelling. On the supply side we would expect the supply curve to exhibit the normal upwards sloping relationship between price and quantity. That is, there will be some landlords who are willing to rent their houses at a low rent, whereas others will only consider it if rental income is to be relatively high. Thus, in Fig. 4.6 we see that an equilibrium rent of *Pm* is achieved and *Qm* houses are rented out. It should be realized by now that such an equilibrium will not necessarily remain stable as either, or both, of the curves are likely to shift over time. For example demand could increase for such properties if there was an increase in student numbers at a local college or university. Or, on the supply side, supply could be reduced if much of this kind of housing was demolished to make way for an inner city renewal scheme, or if a change in taxation made rental income less attractive to landlords for example.

However, with our initial equilibrium rent of *Pm*, the local government, or city council, may feel that this is an unjustly high rent for the less well off communities who usually live in such accommodation. In such a situation the concerned authorities could impose a **maximum price control**, or **price ceiling**, on such dwellings. For example, it could decree that landlords were not allowed to charge more than a rent of *Pc* as shown in Fig. 4.7. This price ceiling would legally prevent the price (rents) rising above

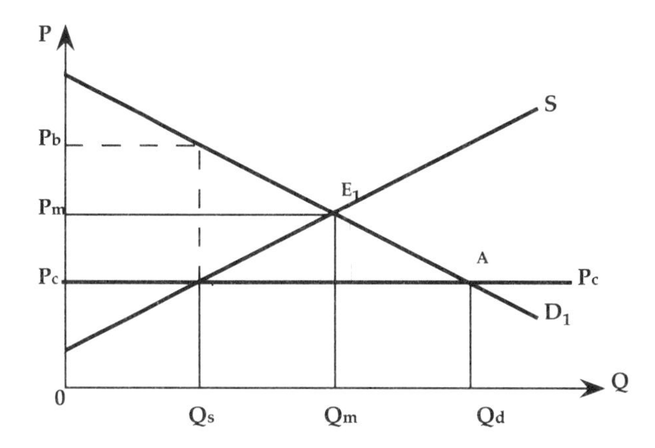

Figure 4.7. *Rental property and the imposition of a rent (price) ceiling*

Pc to the market price of *Pm*. Obviously such a maximum price control is only sensible and effective if the maximum price is set below the ruling equilibrium price. If it were to be set above the equilibrium price, prices would simply adjust back to the market clearing level, as they are free to move as long as they do not *exceed* the price ceiling. That is, prices are now only free to move at or below the artificially imposed ceiling.

After the imposition of such a policy the council may initially feel that it has solved the problem of high rents for the local community as rents have been formally lowered. However, by looking at Fig. 4.7 we can see that we now have a situation where quantity demanded has increased, due to the increased affordability of rented accommodation, from *Qm* to *Qd* as rents have decreased from *Pm* to *Pc*. On the supply side, however, there is a tendency for less accommodation to be supplied as fewer landlords are willing to rent their houses for such a low rent. Thus, eventually, supply could contract to *Qs* from *Qm*, giving us a net situation of excess demand equal to *QsQd*. Such a drop in supply, however, may not be immediate as existing tenants will normally have some form of short-term *security of tenure*. That is, landlords could not throw people out on to the street as soon as the policy was imposed as they would have to provide reasonable and adequate warning. Therefore, it would appear that a maximum price control policy by itself produces quite substantial negative effects as detailed below:

1 Although the local authority has succeeded in decreasing the rental price for the dwellings from *Pm* to *Pc*, this will result in a situation where less accommodation is available than was previously the case. Note that the amount of accommodation available on the market has dropped from *Qm* to *Qs*.

2 To exacerbate the shortfall in (1) above, the level of demand has increased from *Qm* to *Qd*.

3 Low rental incomes could mean that landlords spend less on their properties in terms of repair and maintenance, and shelve any plans for improvement. Such actions would obviously lead to a deterioration of the quality of such housing stock.

4 This situation could lead to an **illegal market** whereby people would 'secretly' pay the landlord a price above the official price ceiling in order to obtain one of the limited number of houses now available. In such circumstances, only the relatively well off could afford to rent a house as the illegal market price (rent) would be as high as *Pb* if there were only Qs houses available to rent. Alternatively, another undesirable outcome could be over crowding as many start to live in the house to share the burden of the high rent.

Therefore, as can be seen from the analysis above, we need to rectify this situation of excess demand by introducing *additional policies* in *conjunction* with the initial price control. Examples of such policies are:

1 An attempt could be made to *shift the supply curve to the right*. Ideally this could be done in such a way so as to ensure that the new supply curve cuts the original demand curve at point *A* in Fig. 4.7, where a new equilibrium would now be achievable. This could be done if the public sector built and provided more rentable dwellings for example. However, apart from the time that it would take to construct such properties, it will require a substantial initial capital outlay by the public sector that may be beyond the budgets of many local authorities. Although, in the long run, costs would only be with respect to ongoing management, and the repair and maintenance of such properties. Figure 4.8 shows the new supply curve *S2* passing through point *A*.

2 We could attempt to *encourage more to be supplied given the existing supply conditions* as represented by the original supply curve S_1 in Figs 4.7 and 4.8. That is, we must encourage the suppliers of rented accommodation, (landlords) to supply more dwellings at this lower rent of *Pc*. Specifically *Qs* properties need to be supplied. This could be achieved by offering the landlords a **subsidy** equal to *PcPs* as shown on Fig. 4.8. However, the problem with this policy is that it 'lumbers' the city council with substantial on-going costs of subsidization as represented by the area *PcPsBA*. Arguably, however, this subsidy could be reduced to a **differential subsidy** shown by the area *ABC* in Fig. 4.8. The theory behind this is the observation that some landlords are willing to accept a lower rental than others. For example, the upwards slope of the supply curve implies that the landlord represented by *Qs* is quite happy with a rent of *Pc*, whereas it is only the landlord represented by *Qd* who insists upon a rent of *Ps* before supplying their property on to the market. Therefore, in theory, different landlords could be given different subsidies reflecting their own individual situation. In reality, however, this would probably be impossible to administer as few landlords would admit that they were

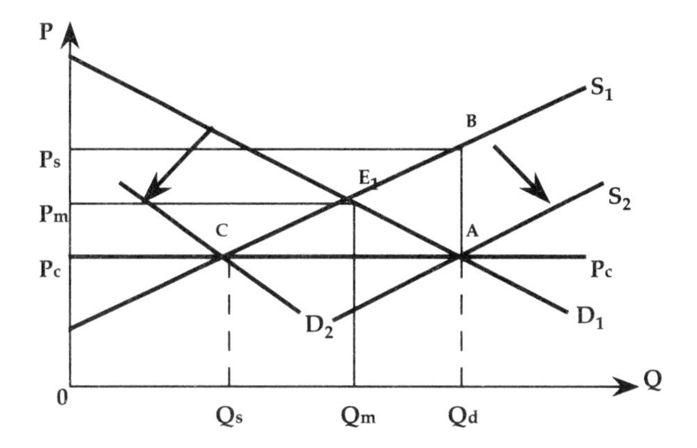

Figure 4.8. *Rent control and additional policies*

happy with a rent below *Ps* as they would all wish to receive the maximum subsidy available.

3 Alternatively, we could tackle the situation by attempting to *reduce demand in this market*. Ideally we would hope that demand could be shifted down to the left so that it could cut the original supply curve (S_1) at point *C* in Fig. 4.8 where a new equilibrium could be established. This could be achieved by encouraging people from this market to move to another market. For example, the government could encourage people to buy their own home and thus enter the owner occupied market by moving out of the rented sector. To stimulate such a shift in demand central government could lower interest rates, or increase (or introduce) mortgage interest tax relief for example.

4 If the initial aim was to reduce the rent for existing tenants from *Pm* to *Pc*, you could make the tenants initially pay the full rent of *Pm*, but give them a **subsidy** (rent rebate) equal to *PmPc*. A problem with this policy is that in keeping the cost of rental accommodation down it is likely to encourage a further increase in future demand, and subsequently therefore, an increase in the number of applicants wishing to receive a rent rebate. This obviously represents a substantial on-going cost for the council. Moreover, as demand rises this may lead to a widening gap between the official rent ceiling (*Pc*) and the potential market rent. Such an outcome is due to the upwards sloping nature of the supply curve and the shifting of the demand curve to the right.

5 Finally, after imposing the price ceiling we could enforce *strict long run security of tenure for existing tenants*. Therefore, despite the receipt of a low rental, landlords could not evict their tenants. However, as mentioned earlier, this is likely to lead to a running down of the physical condition of such properties as landlords cut costs in an attempt to re-achieve their original profit margin.

Example 2: Price control on building materials

One can imagine a situation where a developing nation's government is concerned that in order to facilitate rapidly rising economic growth, more building development needs to take place so that sufficient factories, warehouses, and so on are available for the smooth running of the economy. In order to achieve this goal, the government may feel that it needs to encourage existing building firms to build more, and for new firms to join the market, so as to increase the supply of buildings. One way that the government could attempt to achieve such a goal would be to lower the firms' costs and thus increase their potential profitability.

In order to reduce costs, the government could lower taxes paid by building firms. However, as this would adversely effect revenue to the government they may alternatively attempt to reduce the costs facing building firms by placing a price ceiling on essential building materials, such as cement. As with the case of rented accommodation above, we shall see that such a policy by itself will lead to potentially detrimental outcomes which need to be reduced, or rectified, via additional policies introduced in conjunction with the initial price ceiling. (It should be noted that governments who adopt a centrally planned system of economic administration by regulating prices in the economy, may inadvertently place a price ceiling on goods if prices are not reviewed and adjusted regularly enough.)

By using Fig. 4.7 again, but now for a new example of *the market for cement*, we can see that the immediate impact of the price ceiling policy was to lower the price of cement from *Pm* to *Pc*, and to increase the demand for it from *Qm* to *Qd* as desired. However, the suppliers of cement will now only wish to produce *Qs* cement as more production would not be profitable given the new conditions of a regulated price. As such, marginal cement production would be shut down leaving us with an excess demand of *QsQd*. If this situation was allowed to persist an illegal market would be likely to arise whereby builders have to offer suppliers higher than normal prices (up to *Pb* in Fig. 4.7) so as to obtain some of the limited supply. Moreover, such a situation is often exacerbated as people hoard cement in the hope that further artificial shortages will drive prices up even further so that they can make even larger profits on the sale of the good. Thus, less building work is likely to occur as an essential raw material becomes very costly and largely unattainable. In an attempt to rectify this situation the government could introduce several additional policies to support the initial price control. The following discussion concerning such additional policies should be viewed in conjunction with Fig. 4.8.

1 *Encourage an increase in supply* so that the original supply of cement as represented by the supply curve S_1 shifts to the right to S_2. This could be attempted in a variety of ways:

(i) Allowing imports of the material. The problem with this policy however, is that it uses the countries valuable foreign exchange, and could encourage a dependency on imports.

(ii) The government could try to attract more firms to set up in the production of cement. However, this is likely to be difficult as few firms would find joining the industry an attractive proposition when the price of the good is so low. Therefore, such a policy may have to be promoted by offering firms subsidies or tax incentives, both of which would detract from government revenues.
(iii) A strongly interventionist government may decide to nationalize the industry and enforce greater output. Unfortunately, however, many publicly controlled industries throughout the world have had a reputation of poor organization leading to higher production costs and even lower output in terms of both quality and quantity.
(iv) The government could either finance research, or provide research facilities, to find cost cutting methods of production, or other productivity improvements for the existing firm(s) in the industry. This would have the effect of enhancing the producers' profits thus encouraging higher output.

2 The government could *encourage the existing supplier(s) to supply more by offering them a subsidy* equal to *PcPs*. Alternatively, it may be possible to administer a 'differential subsidy' that could be distributed to the firm if it had different plants with different circumstances which each required different incentives to make them profitable. Such a differential subsidy would be represented by area *ABC*.

3 Alternatively, *the demand for the good could be reduced* preferably so that the demand curve shifted to the left from D_1 to D_2. This could be achieved by trying to reduce the reliance upon cement by changing the method and type of construction. For example more use of different materials, say wood, or the use of more prefabricated components are both possibilities.

4 The government could directly *subsidize the building firms* themselves for every bag of cement used. Initially this would require a subsidy equal to *PcPm*, although this figure could well increase as demand for the product increased due to its effective price decrease.

5 Finally, if supply had fallen to *Qs*, the government could *introduce a rationing system* so as to equitably distribute the limited supply. However, such systems are often expensive to introduce, and hard to effectively police. Moreover, in this instance, it defeats the original objective of encouraging more supply and more building work.

(2) Minimum price controls (price floors)

A minimum price control (price floor) is a situation where governments decree that people cannot charge, or offer, below a certain official minimum price. Therefore, such policies are initiated when the government believes that the price determined by the market is too low. Thus, in such an case, public intervention is thought to be desirable so as to enforce a

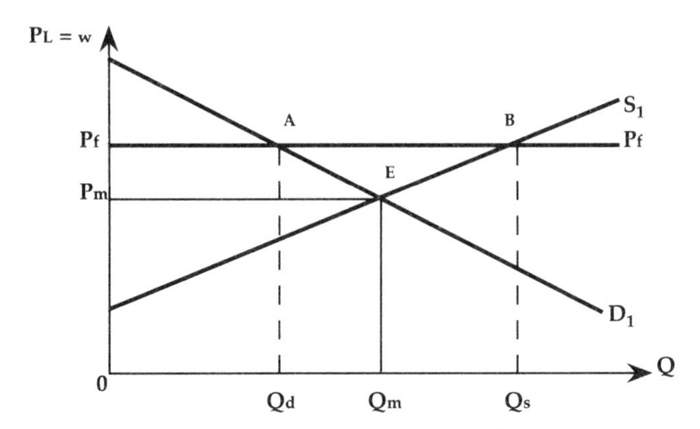

Figure 4.9. *Minimum wage legislation in the labour market*

higher price greater than the market equilibrium. For example, we can see this in many countries in the form of *minimum wage legislation* for certain types of labour. That is, governments enforce a legal minimum wage that must be paid by employers. To have the desired effect of increasing wages such a minimum wage, or *wage floor* as it is sometimes known, must be set at a level greater than the market wage. This is shown in Fig. 4.9 where the wage floor is represented by *Pf*.

Fig. 4.9 shows that if the market were left to itself and operated freely workers (for example, manual, building site labour) would receive a wage (*w*) of *Pm*, and *Qm* people would be employed. However, if a wage floor, *Pf*, was imposed for such labour, wages would have to increase to this new, higher level. Such increased wages are likely to attract more people into the industry due to the higher rewards obtainable. People that joined the industry could be either those who are presently unemployed or those who are working in another industry which now pays less. Given the current supply curve S_1, which represents the potential supply of construction site labour in our example, *Qs*, rather than *Qm*, workers would now offer their services at this higher wage. However, with such high wages employers are likely to seek ways of laying some labour off so as to reduce the rising costs of employment. A reduction in the labour force may be possible as it becomes more economic to replace some labour with more capital intensive methods of production, or firms insist on higher productivity levels from the remaining workers. It is also possible that not only will firms contract by laying off workers, but some weaker firms could go out of business altogether if they fail to cover these increased costs. Thus we are left with a situation of an excess supply of labour equal to *QdQs*. In other words we have *unemployment*. Therefore, just as with price ceilings, we need to introduce additional policies in order to ensure the success of our initial objective. Such accompanying policies could be:

1 To encourage a reduction in the supply of labour so that the supply curve (S_1 in Fig. 4.9) shifts to the left causing it ideally to cut the demand

curve at point *A*. This could be attempted by offering government retraining schemes to retrain the workers to work in another industry.

2 Encourage existing employers to employ more workers by giving the employer an employment subsidy. The ideal result of this would be to increase the demand for labour so that the new demand curve cut the supply curve at point *B*. A subsidy of *PmPf* would be needed to achieve this.

The cost of such policies would be largely determined by the relative elasticities of the demand and supply curves. For example, the demand curve for labour could be made perfectly inelastic from *Pm* to *Pf* if workers agreed to increase their productivity in exchange for the higher wages. In such a situation nobody would be made redundant as a result of the initial policy. If this was the case, supplementary policies would not be needed either.

CONCLUSION

It should now be appreciated that market analysis is both a useful and versatile technique available to you for the understanding, and the prediction, of the workings of many aspects of the built environment. In fact, it would be a very useful exercise for you to apply these concepts to a related example of your own. For instance try to explain how many estate agents there will be in an economy, and what their wages are likely to be given a buoyant, or depressed property market; or explain why certain building materials are used in preference to others – just to name two possibilities. Remember that the examples selected in this text have been used merely for the purposes of illustration, and are by no means an exhaustive range of applications. However, it is important to realize that in some instances the market fails to produce our desired objectives due to the existence of 'public goods' and 'externalities' for example. Such situations are discussed at length in Chapter 15.

PART 2

The 'theory of the firm' in the built environment

This section is specifically designed to show how a variety of key economic concepts can be used to promote an understanding of the running and prosperity of firms involved in the built environment. The main examples are drawn from the construction industry itself and its immediate suppliers. However, a useful exercise to undertake once you have read and understood this section, would be to see how well you could apply these ideas to a firm providing services such as a building surveying practice, or an estate agent for example. It can be done! Furthermore, as you progress through the section, try to think of the implications of these concepts upon directly related factors such as the design of buildings, the materials used in construction, the rate of building, and so on.

Many may feel that the subject matter of this section is highly theoretical. However, what follows are simply statements of logic which *are* applied in real life. Such application is often not realized by people in practice as they may not necessarily be aware of the fact that they *are* behaving in a way described and predicted by the theory. This can simply be due to the fact that they are not aware of, or are not conversant with, the strict terminology of the subject of economics. Furthermore, those firms who do not behave in the ideal manner suggested by theory may indeed improve their performance with a greater understanding of the application of such relevant theory. For example, many a building firm has presented itself with unnecessary problems as they have reacted in the wrong way to changes in the market, or they have failed to adequately predict future changes in demand which may have been forecasted with an understanding of the mechanisms described by the theory. In any case, remember that theory should never be divorced from reality as it is merely an attempt to help to explain 'what happens', and to help to predict 'what will happen'.

Chapter 5, in this section will examine the theory of costs and revenues and how this can be applied to the immediate economic management of a construction firm. Chapter 6 then expands upon this analysis to look at the firm's long-run prospects. In Chapter 7, we analyse the industrial structure of the building industry, and see how this can be better understood with some knowledge of the theories of the firm. To help us achieve this task we will examine two diverse industrial structures that approximate certain sectors of the construction industry and its suppliers, namely the theories of (perfect) competition and monopoly. Finally, Chapter 8 examines the construction industry in more depth and contemporary issues related to it.

5

The theory of costs and revenues, and its application to construction firms in the private sector in influencing their output decisions

As the title suggests, this chapter is primarily concerned with the question: How much should a firm produce? Therefore, if we are looking at a firm of *house builders* for example, the theory that now follows should give an indication to that firm on how many developments to initiate and continue under varying market conditions. However, although the principles hold here, it must be said at the outset that the following analysis is often more complicated in reality as market conditions rapidly change. Continual market fluctuations require an on going up-date of calculations and analysis by the firm's decision makers if their firm is to be successful. Moreover, the following analysis concerns a firm facing a downwardly sloping demand curve, yet some firms may be so insignificant in terms of overall industry output in an area that they become 'price takers' and are subject to a completely inelastic demand curve. For a description of such instances please refer to the theory of 'perfect competition' and its application (p. 105).

As a logical starting point we shall assume that the rational firm in the **private sector** wishes to **maximize profits**, whereby profits can simply be calculated by subtracting all of the firm's costs (total costs) away from its total receipts (total revenue). This can be written as:

$$\Pi = TR - TC$$

where: Π = profit;

TR = total Revenue;
TC = total Costs.

In fact, it is hardly surprising to find that, in reality, profit maximization does seem to be the aim of most private firms, although empirical evidence does show that some firms are more interested in **maximizing sales** rather than maximizing profits. It is worth mentioning that the aim of maximizing sales may be due to mere 'company ego', or due to long-run ambitions of future profitability by initially flooding the market with a firm's product, (so as to make a name and reputation for it as a leading brand), before following purely profit maximizing procedures. It is useful to bear these two aims in mind so as to appreciate that they are two very different goals.

Although, at first sight, one may think that the more the firm produces the more profits it will make, this is normally far from the case as shall now be seen. Also it should be noted that firms in the **public sector** may have other goals such as the **maximization of social welfare**.

To find out how much a firm should produce in order to maximize its profits we need to examine the potential revenues that the firm can receive and the costs that it may face. Importantly we must recognize the fact that anticipated costs and revenues can change over time as the market, in this case the housing market, changes. In the following analysis try to further your understanding, and your ability to apply economics, by keeping in mind how the concepts and processes mentioned could also be used to examine a variety of firms in different situations such as:

1 a building firm (continuing the house builder example from above);
2 a firm producing building components;
3 a firm of surveyors or architects.

(A) Revenues of the firm

Under this title we shall examine the key areas of the firms potential revenue. At first many may feel that there is a lot to learn here and that there are many new terms and concepts to understand. However, you will soon appreciate that such a fear is largely unfounded inasmuch as all the ideas below are inter-related. Furthermore, if you take time to think about the meaning of the terms themselves they are really quite straightforward and self-explanatory.

(1) Total revenue (*TR*)

Total revenue (*TR*) is simply the total amount of money that a firm receives from selling its good or service. Thus, total revenue received by the firm is the obtained price of the good or service multiplied by the particular level of output that the firm manages to sell on to the market. Therefore, total revenue received by the house builder can be calculated from the actual sale price of their houses on completion of contract multiplied by the number of houses that they actually sell. We can express this as:

$$TR = P \times Q$$

where: p = price,
q = quantity.

As total revenue is concerned with a relationship between **price** and **quantity** we can estimate the potential revenue facing the firm by examining the demand curve for that firm's products at the present time, or at least the demand curve that the firm is likely to face when its product or service comes on to the market. As already implied, it is important to appreciate that such a process should be under continual review as market demand can change over time.

Therefore, let us go back to the example of the house building firm and imagine that they have estimated that the likely demand for new 'starter homes' in a town is represented by the demand curve in Fig. 5.1. As suggested above, this demand curve may need to be an estimate of the level of demand in the future as it will take some time to get the product (houses) on to the market. For examples, time will be spent assessing the market in the first instance, obtaining planning permission, finding suitable building land, preparing the site, organizing production, and so on.

The perceived demand schedule in Fig. 5.1 suggests that the firm expects that it would sell none of its starter homes if it set the price too high, say at a price of P_0, and therefore no revenue would be received. This result is because, nobody looking for a starter home would be willing to pay such a high price as normally cheaper alternatives would be available from the 'second hand market', or people would be attracted to the 'rented sector'. On the other hand, if the firm thought of restricting output on the site to ten houses (Q_1), for example, it would estimate from the demand schedule that it could sell the dwellings at the relatively high price of P_1. This high price would be possible only if no near substitutes were available at a lower price that potential house purchasers may buy instead. If there were no near substitutes available, high prices would prevail because it would then be likely that there would be fierce consumer competition for the small number of houses available on the market (refer back to Chapter 1 on the determination of market equilibrium). However, note from the area $0P_1AQ_1$ in Fig. 5.1, that although a high selling price would be achieved, overall total revenue would be low as so few units would have been produced and sold.

Alternatively, by constructing fifty houses (Q_2), the level of competition between potential purchasers is unlikely to be as great as there is more

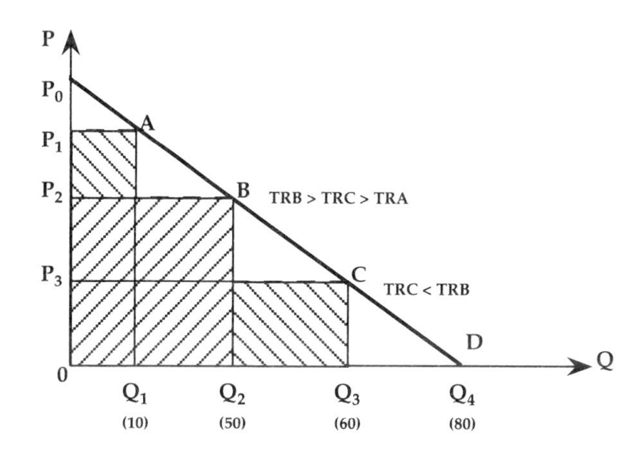

Figure 5.1. *The demand curve and total revenue*

choice, and therefore the market will only clear at a lower equilibrium price of P_2. Thus, despite a lower selling price, more houses could be sold and a greater total revenue received. That is, the area of total revenue at the lower price of P_2, given by the area $0P_2BQ_2$, is larger than the area of total revenue gained at the higher price of *P1* of OP_1AQ_1.

Building more though, will not necessarily increase total revenue further. This is simply because if too many houses 'flood' on to the market, the supply curve for such housing will shift to the right and prices will be driven down. In other words, there would be an increased level of competition between the sellers of such houses, both old and new, to attract the limited number of people interested in starter homes as represented by the demand curve. Therefore, if sixty houses were built, the price would be driven down to P_3. Such a substantial drop in price would lead to low revenues despite the large number of houses sold. Note that the area $0P_3CQ_3$ representing total revenue received from the sale of sixty houses is less than area $0P_2BQ_2$ which represents the monies received from the sale of just fifty homes. In fact if too many houses are built there may be nobody interested in buying them if there is not sufficient demand in the area, and some buildings would remain unoccupied. Figure 5.1 suggests that in the case of our limited example, at the theoretical extreme, even if the houses were free only eighty would be occupied as there would be a finite number of people looking for such a house in that area. Moreover, a parallel consideration is that if a site is too densely built on it becomes less attractive to potential purchasers who may decide to look elsewhere for a better planned, more spacious development as a place to live.

Therefore, if the firm is faced with a downwardly sloping demand curve, as is likely to be the case, one will find that total revenue initially rises with increased sales, yet it will reach a peak and decline again. This conclusion and its supporting information is transferred from Fig. 5.1 to Fig. 5.2. The areas under the demand curve in Fig. 5.1 give us the figures for total revenue that enable us to plot the 'total revenue curve' in Fig. 5.2.

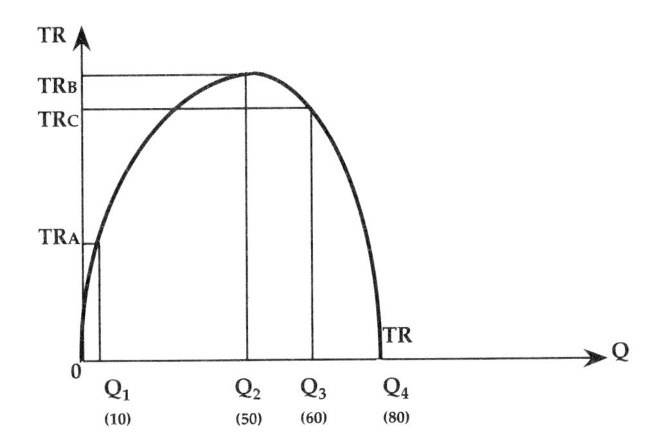

Figure 5.2. *The total revenue curve*

Before we leave this introductory discussion on total revenue please note particularly the following statement: *Total revenue is not the same as profit*, it is simply the total monies received by the firm. Therefore, at this stage, the firm still does not know how many houses to build or how much profit it will make. To discover comparative profitability at each level of potential output the firm also needs to obtain detailed information concerning its costs. Moreover, a slightly more detailed analysis of revenues can be useful to the firm as well.

(2) Marginal revenue (*MR*)

The concept of 'marginal revenue' (*MR*) follows on directly from above, and is simply obtained by measuring the change in total revenue caused by selling an additional unit of output. Thus:

$$MR = dTR$$

where d = a change in.

If nothing is sold, for example, total revenue would be zero. Then, initially, as you sell some houses, total revenue obviously increases, and the change in total revenue is positive. However, with the given demand conditions, the addition to total revenue being created by more houses being sold will become less and less as prices have to be lowered to sell the greater level of output. In fact, one would reach the stage where the price of the good (houses) had to be dropped so much in order to sell all available output, that additions to total revenue (marginal revenue) become negative. Once this point has been reached the total revenue curve is at its peak and will now begin to decline, as can be seen in Figs. 5.1, 5.2 and 5.3. These diagrams show that if the firm is faced with a downwards sloping demand curve, the more it sells the lower will be the selling price of its

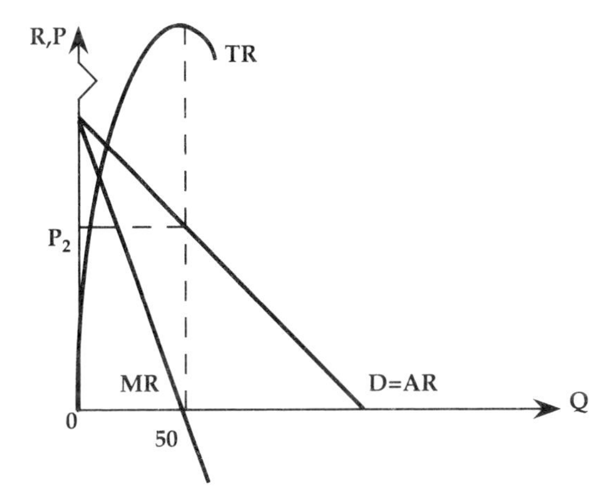

Figure 5.3. *Marginal revenue derived from total revenue*

output, and therefore its marginal revenue curve will also be downward sloping. For example, it was seen that in order to sell sixty houses instead of fifty the price would have to be dropped from P_2 to P_3 to ensure that all the houses are sold in the current market conditions. In other words, marginal revenue declines as output increases.

(3) Average revenue (*AR*)

Average revenue is simply how much the firm receives on average. Thus, mathematically, average revenue is total revenue divided by the current level of output sold and can be expressed as:

$$AR = \frac{TR}{Q} = \frac{P \times Q}{Q} = P$$

From this it can be seen that average revenue is the same as price at any given level of output. It therefore follows that the average revenue curve is the demand curve as it is the demand curve which indicates at what price the goods (houses) will be sold. Thus:

$$D = AR$$

To understand the distinction between average revenue and marginal revenue note that the additions to total revenue (marginal revenue) are zero at fifty houses, yet average revenue is still positive at P_2. This is because although houses are still being sold, and revenue is still being received, the price is so low that total revenue would have been more if the firm had tried to sell fewer houses at a higher price.

Conclusions on revenues

The above analysis is straightforward, but it must be remembered that it all hinges on an accurate forecast of the exact location of the demand curve. Moreover, remember that such demand can be quite volatile in many markets. For example housing demand could easily be affected by changes in government macroeconomic policy and in particular interest rate policy, demographic change, etc.

(B) Costs facing the firm

For a firm to find its optimal level of output, with respect to making the maximum possible profit, it must also discover what costs it faces. In order to do this the firm needs to work out the cost of producing every level of output. Once it has this information, it can be compared with the corresponding level of revenue so as to ascertain the optimum level of profit. Just as with revenues, it must be emphasized that the assessment of the firm's costs should be a continual one as costs often change reflecting general inflation, or sudden supply shocks to the economy, for example. Thus, when forecasting the future likely costs of a firm the figures are obviously estimates and subject to a degree of error. Furthermore, when considering costs, we need to firstly analyse the short run situation facing the firm, before going on to examine its potential long-run situation. This is an important distinction because it is only in the short run that a firm is faced with a situation where certain factors are fixed. For example, if an entrepreneur purchases, or rents, a small factory, or office, in order to produce a good or service, he or she is committed to this accommodation, at least in the short run. That is, if the firm then wished to increase output beyond the capacity of the present building, it would take time for extensions to be built, or new premises to be found. Therefore, in the short run, only variable factors can be adjusted such as the number of people employed. The long run, however, is a period of time that is sufficient for the entrepreneur to vary all of the firm's inputs so that the optimum size of operation can be reached. The time periods involved between the short run and the long run will depend upon the industry in question as, for example, the opening of a new branch office for an estate agent is obviously easier and quicker than the setting up of a new brick works or cement factory.

Due to this distinction between the short run and the long run we will now look at how costs vary in each time period, while, long-run cost will be covered in Chapter 6.

Short-run costs

(1) Fixed costs (*FC*)

Because some factors are fixed in the short run, the firm will incur fixed costs that cover these factors regardless of the level of output. These are usually referred to as 'overheads'. Examples of such costs are:

1 rent of premises whether a shop, factory, or office;
2 rent of any plant and equipment;
3 business rates payable to the local authority;
4 insurance;
5 the cost of indirect labour not directly involved in production, such as essential management and administrative staff;
6 interest charges on borrowed monies.

Thus, whether our house builders build a hundred houses or no houses at all, they would still have to meet these fixed costs. The situation would be the same for a firm of surveyors, for example, whether they undertook any surveys or not. Such costs can only be changed in the long run.

(2) Variable costs (*VC*)

As the name suggests, variable costs are the costs of inputs which the entrepreneur can alter (vary) depending upon the level of output of the firm.

Examples of such costs are:

1 the wages of productive labour;
2 the quantities of raw materials or supplies used;
3 the cost of transport;
4 energy costs of production.

Therefore, variable costs are dependent upon the level of output of the firm. For example, as more houses are built by the house building firm it will need to employ more staff, purchase more raw materials, etc.

(3) Total costs (TC)

From the above it can be seen that in order to produce a good or service the firm will be confronted by both fixed costs and variable costs. These two cost categories added together represent the total costs facing the firm. This can be expressed as:

$$TC = FC + VC$$

where TC = total costs;
FC = fixed costs;
VC = variable costs.

(4) Average costs (AC)

Using the above information we can now find out what costs will be on average depending upon the level of output. Specifically we can find out average total costs by dividing total costs by the particular level of output that we are interested in. Likewise both average variable costs and average total costs can also be found in this manner:

$$ATC = \frac{TC}{Q} \qquad AVC = \frac{VC}{Q} \qquad AFC = \frac{FC}{Q}$$

The minimum point on the average total cost curve represents the minimum cost of production given the firm's existing size and short-run fixed factors. This is not, however, the point of profit maximization as such a point also depends upon relative revenues (see the profit maximizing rule below)

(5) Marginal costs (MC)

Marginal costs are simply the additional costs of producing one more unit of output. Thus, marginal cost can be expressed as:

$$MC = dTC$$

where d = change in.

Figure 5.4 illustrates total costs, variable costs, and fixed costs and shows how they behave as output changes. Firstly, it can be seen that fixed costs

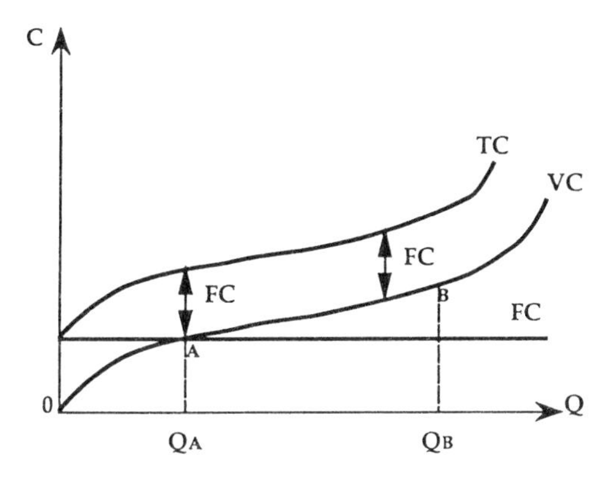

Figure 5.4. *Total costs, fixed costs and variable costs*

remain constant irrespective of output. On the other hand, variable costs, such as the cost of hiring labour, initially tend to increase rapidly as the first few people hired by the firm, for example, are likely to be too few to be fully efficient. Therefore, the firm is paying wages in exchange for relatively low output. This situation is represented by the steep gradient *0A* on the variable cost curve. After this a stage could be reached whereby additional employees have sufficient numbers of support staff such as technicians and secretaries, so that each extra individual employed could work to their full potential and thus add greatly to the output of the firm. Therefore, although more wages are being paid, output is increasing at an even greater rate. This situation is represented by the shallow gradient *AB* on the variable cost curve. This portion of the curve shows great increases in output, yet relatively small increases in the variable cost bill. However, considering that we have fixed factors in the short run, the **law of diminishing returns** is likely to eventually set in as we increase output. That is, in order to increase output yet further (beyond *QB* in our example) overtime payments may have to be made, and more staff may have to be taken on. Under such circumstances wage costs will increase, yet additions to output may be quite marginal as new employees do not have sufficient access to support staff and equipment and therefore their productivity is low. Thus, variable costs would now rise rapidly without a corresponding increase in output as shown by the variable cost curve beyond point *B*. In a frantic effort to produce even more, under the given short-run conditions, the firm could reach the stage where it had employed so many people that the workplace would become overcrowded and there would not be enough room for the new employees to work even if they wanted to. Here the wage bill would rise without any corresponding increase in output. In fact, at the theoretical extreme, yet more workers could mean that new workers are actually getting in the way and damaging the productivity of existing labour. These problems determine the shape of the variable cost curve, and as we add this to the constant fixed costs we find that the total cost curve is the same shape as the variable cost curve yet at a higher level. The distance between variable costs and total costs being represented by the level of fixed costs.

This relationship between labour productivity and output with fixed factors of production is shown in Fig. 5.5. In this example all workers up to worker L_1 would contribute in an increasingly positive manner to the firms output as the organization and specialization of labour allows for greater productive efficiency. However, if more than L_1 workers are employed, with limited space and equipment, the marginal productivity of additional workers will fall. In fact, by employing more than L_2 employees additional workers would have a negative productivity that would actually damage the work of existing employees.

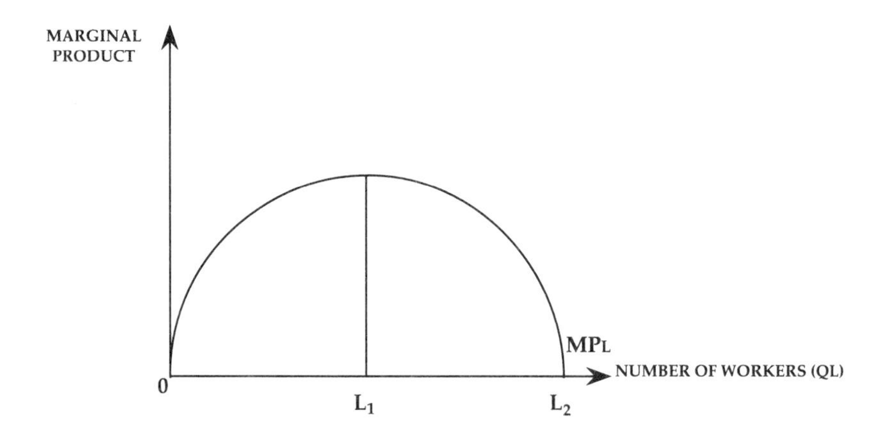

Figure 5.5. *Marginal productivity of labour*

To understand these points yet further, imagine a small firm of house builders which presently occupies a couple of offices in a building. The firm will obviously incur fixed costs, of the type described above, irrespective of its output. That is, in the short run, even in the middle of a recession, if the firm wished to contract, for example, it would take time to change premises and sell off other equipment such as drawing boards and computers. Alternatively, if the firm was building more during a 'housing boom' it may need to employ more administrative and managerial staff at the head office and thus increase its variable costs. Larger offices may be required to accommodate such expansion, yet if this is not possible in the short run the problem of diminishing returns may well be experienced as pressures on space mean that decisions and work is delayed.

By using logic, or a numerical example, we would find that average total costs, average variable costs, and marginal costs, all decrease at first, representing the increasing efficiency of the firm as it expands output, and then begin to increase due to the law of diminishing returns. Average fixed costs will constantly decline as we divide an increasing level of output into the constant cost. These costs have been plotted in Fig. 5.6.

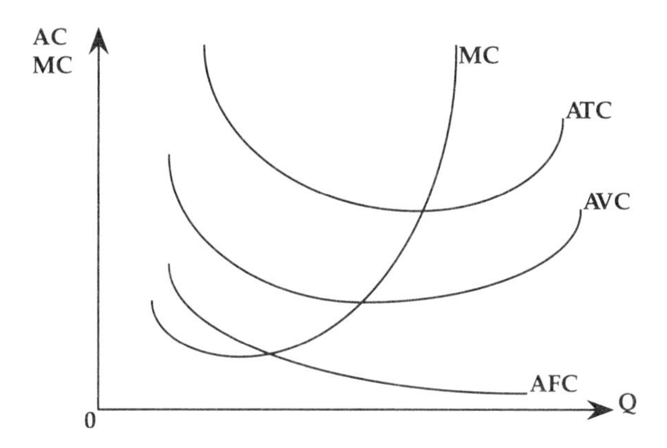

Figure 5.6. *Average costs and marginal costs*

A note on 'normal profit'

You will find that the term **normal profit** is frequently referred to in economics and in latter parts of this book (see Chapter 7). Normal profit is the amount of money that entrepreneurs need to make from the firm in order to keep themselves in the business. Such an income needs to be at least equal to the opportunity cost of the entrepreneur being involved in the firm. For example, if the entrepreneurs have set up their own building firm after being employed by another firm, they would need a reward at least equal to their old wage to keep them in their own business. Thus, such normal profit is considered as a cost of production as it is a necessary payment for the organization of the firm.

The profit maximizing role

Once one has accumulated information on both costs and revenues one can find out how much profit the firm is expected to make at each level of output. By looking at Fig. 5.7 one can see that the greatest level of profit is represented by the maximum distance between the total revenue curve and the total cost curve (distance *AB*). For example, the level of profits at Q_1 is obviously greater than those at Q_2 as there is less of a differential between total costs and total revenues at Q_2. That is, distance *AB* is greater than distance *CD*. Visually it can be quite difficult to ascertain the exact point of profit maximization in this way as you have to compare relative distances. Therefore, it is useful to be aware that the point of maximum profit will coincide with the point where the marginal cost curve and the marginal revenue curve intersect (point *E*). This gives us the **profit maximizing rule** that states that: In order to maximize profits the firm should produce up to the level where marginal cost equals marginal revenue, or to be more

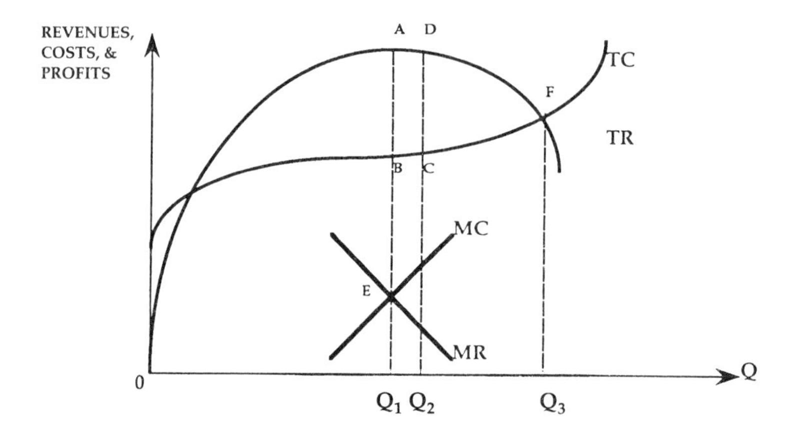

Figure 5.7. *Profit maximization*

precise, they should *produce up to the point where the marginal cost curve cuts the marginal revenue curve from below.* This rule can be explained by the fact that for any level of output less than Q_1 the revenue received from producing each additional unit of output (marginal revenue) is greater than the cost of producing each additional unit of output (marginal cost). Therefore, as long as marginal revenue is greater than marginal cost, profits will be added to as more money is being received per unit of production than is being spent to produce and sell it. However, if we produced more than Q_1, marginal revenue would be less than marginal cost, and as such profits would begin to be eroded away. In fact if the firm continued to increase production up to Q_3 it would deplete all of its profits, and merely 'break even', as it sold units for less than cost price. Note that total costs and total revenues are the same at this point (point *F*). Increasing production beyond Q_3 would mean that the firm's total costs now exceeded its total revenues and an operating loss would be made. It is important to appreciate though that this point of profit maximization (Q_1) could shift over time due to changes on the demand side of the market, as well as possible cost (supply) changes. However, being able to forecast such conditions should enable the house builder to decide upon how many houses to build; the building materials supplier to decide upon the most profitable level of output; and the surveying firm to decide the most profitable level of work to take on.

6

The long-run planning decision: the theory of long-run costs

As a firm develops, it may feel that the market is sufficiently large, or is likely to grow in the near future, to a point where it would make good economic sense for the firm itself to grow. The firm's existing factory, retail outlet, or office, may be too small to cope with such an eventuality, and any increase in output in the existing premises, with its fixed factors, will lead to diminishing returns (see the discussion on short-run costs in Chapter 5). The firm could try to estimate the cost implications of moving into a larger building as part of its long-run planning decision. In order to do this a series of **short-run average total cost** curves (SRATC) could be drawn up representing different potential sizes that the firm could grow to. In examining the growth of the firm in this way it will be found that various factors collectively known as **economies of scale** will tend to cause long-run average total costs to initially decline as the firm becomes larger and produces more. However, if the firm were to become too large, it is at risk of experiencing **diseconomies of scale** which are forces that could lead to rising average total costs. (Both economies of scale and diseconomies of scale will shortly be discussed at length.) Thus, given current cost and market conditions, we would find an **optimum size** or level of output that the firm should grow to. Such a point would be identified by the minimum point on the firm's **long-run average total cost** curve (LRATC).

Therefore, forces are present to give us a 'U-shaped' long-run average total cost function. This curve is made up of all the potential sizes of the firm during its growth. As this curve can be estimated prior to its expansion it is sometimes referred to as the **long-run planning curve**. It should be noted that in the **very long run**, technological change, for example, may alter the position and level of this minimum point. For instance, cost cutting technology could benefit firms of all sizes in the industry and thus lower their short-run average total cost functions. If this was the case, the overall long-run average total cost function, which summarizes these short run functions, would also be lowered. Figure 6.1 shows such a long-run average total cost curve. Here, if only Q_1 were to be produced, the small firm represented by the short-run average total cost curve $SRATC_1$ would still be the optimum sized firm as it could produce this level of output at the lowest cost of C_1. Note that the larger, 'mid-sized' firm could only produce Q_1 at a higher cost of C_2, perhaps due to the under utilization of the larger firm's fixed factors. This *partially* explains why some firms, such as estate agents for example, who need a physical presence in the market

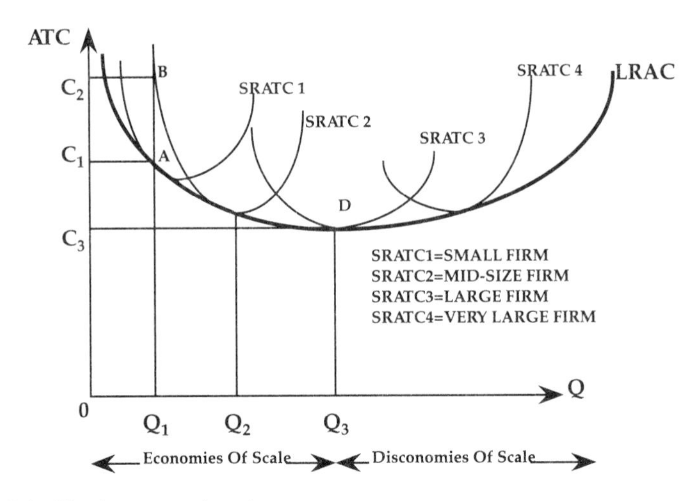

Figure 6.1. *The long-run planning curve*

itself, have many small branch offices rather than just one large one. However, due to diminishing returns, the firm, in its existing state, would have found that increasing output would only lead to rising costs unless it expanded into larger premises. Therefore, if our small firm wished to produce Q_2, for example, it would need to expand into the size of firm as represented by the 'mid-sized' firm. If the business continues to expand, it may be able to take advantage of economies of scale so that it could eventually produce Q_3 at the low cost of C_3, thus being able to 'undercut' the smaller firms.

(A) Economies of scale

'Economies of scale' is a rather cumbersome expression which simply means that cost savings could be made as a firm expands and moves into larger premises. Indeed many may find it easier to refer to as 'economies of size' although this expression is less common. Many economies of scale exist, but only some are directly applicable to certain types of firms as will become obvious as we go through some key examples below.

(1) Raw materials

As a firm becomes larger, and increases its size of production and level of output, it is likely to be able to obtain discounts on the bulk purchase of raw materials and general supplies, thereby decreasing its costs. For example, a large firm of house builders will probably represent a major client for many building supply firms. Thus, in order to retain the house builders' custom, the materials suppliers are likely to offer discounts, better speed of delivery, and so on. Such cost reductions will, by definition, lower the building firms' average total costs. Linked to this idea is the fact that many large firms, for example of surveyors, will receive discounts on office supplies such as stationery for the same reason of bulk purchase as above.

(2) Finance

When financial institutions lend out money they do so at their own risk. Therefore, lending to a larger firm is likely to be seen by the financial institutions as being less risky than lending to a small firm. This is because it is more probable that a large firm has considerable assets, and possibly financial reserves, at its disposal if it had to weather a prolonged recession for example. Such increased security enables the lender to offer the loan at a lower rate of interest than would normally be the case if they were lending to a smaller firm. Small firms are unlikely to have the assets or reserves of a larger firm and thus face a higher likelihood of bankruptcy. That is, lenders would wish to receive a higher reward (rate of interest) to compensate for the greater degree of risk that they take when lending to smaller enterprises. However, not only may the cost of a loan be cheaper for the larger firm, but it is also more likely that it will obtain the loan in the first place due to the higher level of security that it can offer the lender, and there may also be a greater variety of types of finance available to it. Such financial economies are an important economy of scale in the construction industry as most projects do involve large sums of money, with projects ranging from large civil engineering work to extensive refurbishment of existing buildings. Again such savings will tend to lower the average costs of production of the larger firms.

(3) Research

Larger firms are likely to be able to afford expenditure upon research and development, or even have their own 'R&D' departments, whereas the small firm is unlikely to be able to devote resources to this area. Research is useful for a variety of reasons such as improving the product and thus encouraging higher demand (and subsequently more revenue). However, an important aim of research is to find cheaper, and more efficient methods of production, thereby lowering the average costs of production. To further this debate it would be useful to compare the perfectly competitive firm against a monopoly as seen in Chapter 7. The house builder, for example, could undertake research in order to investigate ways of producing more houses at a lower cost, subject to conforming to the constraint of acceptable building standards and working practices.

(4) Specialist equipment

Larger firms tend to have more money to invest in modern, efficient, (perhaps computerized) techniques of production. Such investments are likely to lower the average costs of production. For example a building surveying practice may be able to complete surveys and give clients advice more quickly if they have access to the latest laser equipment or computer aided design facilities (CAD). In achieving faster surveys the firm could reduce its staffing costs as fewer individuals can now complete more work. Similarly, the large construction firm may be able to purchase a computerized site management system so as to improve productivity on site and thus enable them to complete and dispose of the development at a faster rate. Such improvements in productivity could save on costs by reducing the time period over which loan finance is required, and indeed the duration of the hire of subcontractors and plant.

(5) Management

As the firm grows, it should be able to spread some of its managerial and administrative costs over a greater volume of output. For example, even if the firm grows threefold it is still only likely to require one manager, one personnel manager, and one accountant, rather than three of each. Thus, the cost of these employees is spread over the larger level of output of the firm.

However, despite such potential economies of scale, as the firm contemplates further expansion, there may come a point when it will have reached a level of production in a plant size that places it at an optimum (minimum) point on its long-run average cost curve given the current state of technology. After this point diseconomies of scale may set in, and therefore it would not be worth increasing output yet further. This optimum long-run position is represented in Fig. 6.1 by the 'large firms' short-run average total cost curve ($SRATC_3$) giving an optimum level of output of Q_3 at a cost of C_3.

(B) Diseconomies of scale

The term 'diseconomies of scale' encompasses a wide variety of factors that tend to put upwards pressure on average costs if the firm becomes too large. Thus, once such diseconomies of scale set in the long-run average total cost function of the firm rises. Therefore, diseconomies of scale can be seen by the upwards sloping part of the long-run average total cost curve as shown in Fig. 6.1 as the firm produces an output in excess of Q_3. Many potential 'diseconomies' exist and some of the main ones are discussed below.

(1) Inefficient management

Research into very large firms has revealed that inefficient management is one of the main sources of diseconomies of scale. For example, in many cases communication was shown to become progressively more difficult as the firm grew. Because of communication difficulties amongst the firm's personnel the risk of mistakes, uncertainty, and inefficiency are likely to increase. By using the example of the firm of house builders again it can be seen that if the firm has developed rapidly from a small local builder into a large national builder, it may not be able to cope with the difficulties of large scale management, and the co-ordination and organization of many different development sites in different geographical areas.

(2) Specific labour shortages

Specific labour shortages are most likely to occur in industries requiring skilled labour with a certain training. A firm may come up against this problem when it expands as it may require more skilled personnel, such as managers, to effectively organize the increased level of output. In order to attract more of such labour the firm will have to offer incentives which normally take the form of higher wages, or other monetary incentives such as a 'relocation package'. Even if financial inducements are not offered, many firms will pay for the training of some of their staff so that they can obtain the desired skills and necessary qualifications. All of these scenarios will push up the firm's average total costs of production. For example, during a property boom that creates an increased level of activity in the built environment, a firm of surveyors may need to tempt new graduates with high starting salaries and other benefits such as a company car. Similarly, house building booms, and subsequent rapid growth in the housing market, and thus the rate of growth of new house building, have often led to shortages of skilled tradesmen such as plumbers, electricians and carpenters. In these situations house builders have had to pay increased wages and bonuses to such workers in order to attract them to work on their site. The building firm is forced into this situation because if they did not offer higher rewards to such workers they would not attract the people they required and therefore the completion of the development may be delayed. This could make the firm miss out on the boom as it could not sell the completed development at the high prices of the boom.

(3) Limits on existing sources of materials

The existing sources of materials that a firm draws upon may not be adequate to meet the increased demand for them as the firm expands. Therefore, the firm has to develop new sources of materials, or begin to import them from further afield. Both of these solutions are likely to lead to rising costs. For example, imagine a materials supplier to the construction industry such as a brick manufacturer. If this firm has sited its plant near a suitable type of clay for making bricks it will incur low costs of transport in getting the clay to the factory. However, as the original resource runs out, more expensive, deeper extraction will be required, or clay will have to be transported to the plant from further afield, thus increasing costs.

(4) Finding new markets

A firm's medium term growth may be constrained by the size of its market. However, as the firm expands, new markets will have to be sought and this could be very costly. For example, imagine a firm of building surveyors that decides to diversify into a new product area such as estate agency. At the very least, money will have to be spent by the firm upon advertising so as to make a name for itself in its new field of operation.

Conclusion on economies and diseconomies of scale

As can be seen, a variety of circumstances can occur that will put pressure on long-run average total costs to initially decrease as the firm expands, but then increase as the firm becomes too large. However, concerning this analysis, several points should be noted:

1 The above is by no means an exhaustive list. It simply highlights some of the main economies and diseconomies of scale found in most empirical findings of the growth of businesses.

2 Some of the forces mentioned above are only applicable in full to certain types of firm and industry. Although, adaptions or approximations of most of these forces are found in most cases.

3 Some firms will be better than others at exploiting economies of scale and avoiding diseconomies of scale. Such firms are likely to face 'L-shaped' or 'flat bottomed' long-run average total cost curves, rather than the 'U-shaped' curve that the traditional theory predicts. This can be seen in Fig. 6.2 with the last two curves showing the successful avoidance or postponement of diseconomies of scale.

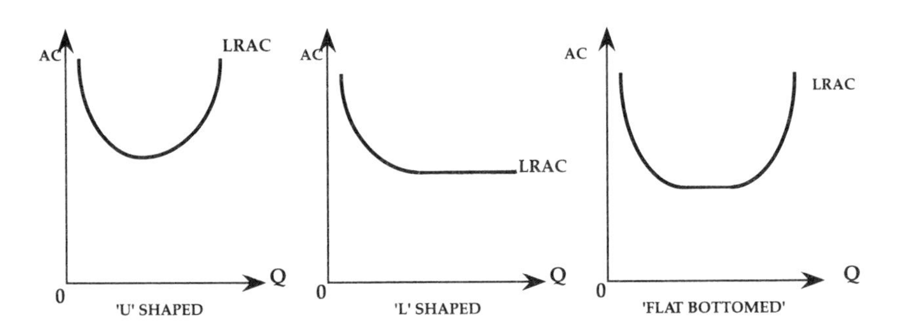

Figure 6.2. *Long-run average cost curves*

— 7 —

Industrial structure and the construction industry (the theory of the firm)

A basic understanding of the fundamental economic forces acting upon any firm, or group of firms, in any industry, helps one to understand an almost limitless range of applied issues such as:

1 how profitable firms are, or are likely to be, in an industry;
2 how a firm's profits can be enhanced;
3 how firms produce their good or service;
4 how many people are, or will be, employed in an industry;
5 explaining regional variations in the prosperity of firms in the same industry;
6 how general economic conditions, such as a recession, can affect the firm;
7 how both local and national government policies can affect the firm;
8 the implications in terms of price and quality, for example to other firms, or consumers, who order products from firms in the industry.

Again this list is certainly not an exhaustive one, but this part of the text will hopefully give you an insight into how economic theories of the firm can be used to address some key practical issues. The following analysis applies the theory to the construction industry and its immediate suppliers. However, these principles can equally be applied to a service profession such as chartered surveying.

As a starting point, the book will identify the main types of industrial structure that one can normally observe in most economies. The first point that should be appreciated is that when attempting to categorize industries, or indeed parts of industries, into forms of industrial structure, one will find that, in most countries, a wide range of industrial structure exists. Generally, one will find a 'spectrum' ranging from firms operating in a highly competitive environment, to those who face little, or no, competition. The exact situation that the firm finds itself in will effect all of the points listed above.

To help us in our task, economics gives us specific terminology that has been devised to identify particular types of industrial structure as can be seen in Fig. 7.1. Figure 7.1 is a 'spectrum' which demonstrates the range of industrial structure from that of **perfect competition** to that of **monopoly**. Specifically it identifies the following possibilities:

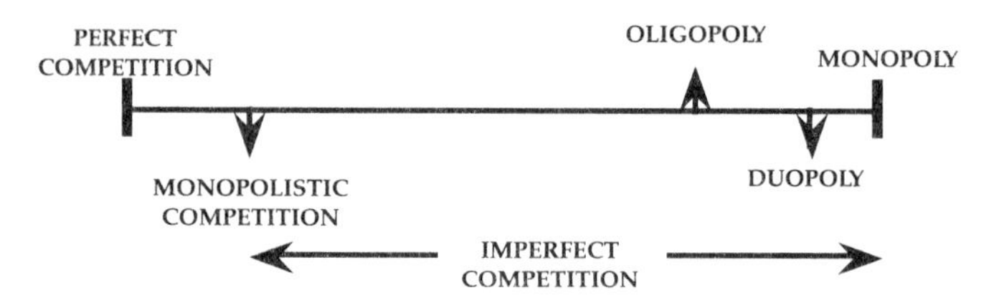

Figure 7.1. *The spectrum of industrial structure*

1 **Perfect Competition**: This is where a large number of fully competitive firms in the industry produce identical goods or services.

2 **Monopolistic Competition**: Here we have a large number of firms in the industry who are able to reduce the level of competition by differentiating their good or service in some way so that their product has a specific, individual identity.

3 **Oligopoly**: An oligopoly exists where several large firms dominate the industry. If these firms collude together to agree upon pricing and output, a **cartel** is said to be formed. Cartels are often thought not to be in the consumers' interests and are outlawed in many countries and in the field of international trade.

4 **Duopoly**: A duopoly is where two large firms dominate the industry.

5 **Monopoly**: A true monopoly will occur if there is only one firm in the industry.

Thus, it can be seen from the above that there are two analytical extremes: very high (perfect) competition on the one side, to no competition (pure monopoly) on the other side. Any industry whose characteristics are different to these will fall somewhere in the middle of this spectrum. For purposes of clarity with respect to using precise terminology, it is important to note that all forms of market structure, except pure perfect competition, can be categorized under the overall heading of **imperfect competition**. Although, in reality, it is unlikely that we will see an industry that is perfectly competitive, or a firm that has a complete monopoly, the study of these situations enables us to understand the behaviour of, and the implications of, firms that approach either extreme. Moreover, by examining these structures one should be able to form an opinion as to whether increased competition, say in the field of estate agency, or in the supply of building materials, or the construction of houses is a good thing or not.

In order to prevent this section of the book becoming an over-lengthy preamble about every conceivable type of industrial structure, I believe that it is sufficient at this level of study, merely to examine the two ends of the spectrum, namely perfect competition and monopoly. As suggested above, it is in this way that one can see the implications facing firms that approach either extreme. However, if you do wish to find out about the other industrial structures in more detail, such as monopolistic competition, there are many 'mainstream' economics texts available on the market, or available in libraries, which would certainly be useful additional reading.

(A) The theory of 'perfect competition' and its application

Many have argued in the past that certain parts of the construction industry have firms, such as the large number of firms in the repair and maintenance sector, which operate in such a highly competitive market that this sector can therefore be seen to approximate the model of perfect competition. Although this is perhaps true in some instances, it is at risk of being a generalization. For example, some small, perhaps remote, communities may only be served by one or two local building firms. There, if repair work is needed on a house, the choice can be, in reality, severely limited. In other words, data may show that there are thousands of firms involved in building repair work, yet it is the distribution and accessibility of these firms which is important. What is, however, a more realistic stance is to firstly select an area where we know that a degree of competition exists. For example if one were to examine a number of house builders developing sites in a region one could try to assess the degree of competition between them, and how well their behaviour approximates the model of perfect competition. If the structure of the industry that these firms operates in does approach the conditions of the model, we can then use it to draw some useful conclusions and answer such points as those raised at the beginning of this part of the book.

Thus, as implied above, to find an industry that truly reflected the model of perfect competition would be nearly impossible as such a large number of restrictive criteria need to hold in order to produce this perfectly competitive state. However, it should be recognized that some industries are so highly competitive that they do indeed approximate the theory. Moreover, it should be noted that it has been the policy of many governments, primarily on the grounds of increasing national efficiency, to encourage a higher rate of competition between firms in an industry. In fact, some politicians have used the theory as the goal of a 'perfect market' and a 'liberalized economy'. Therefore, by examining this model we can begin to see the rationale for, and the implications of, such public objectives.

There now follows an explanation of the assumptions of the model along with a debate with respect to the realism of the assumptions and their application to 'real world' situations in the built environment.

Assumption 1: A large number of firms

For the theory to hold there must be a situation where there are such a large number of firms in competition with one another, that any one firm only produces an insignificant share of total industry output. Thus, any change in output by one firm has little or no effect upon the overall market and the ruling market price. As such, because the firm has to accept the going market price for its good, and cannot influence it in any way, the firm is known as a **price taker**. Diagrammatically this can be seen in Fig. 7.2. Here it is shown that even if the individual firm changed its output from q_1 to q_2, such changes would be so small in terms of the overall market supply that they would not influence the market equilibrium price of P^*, or equilibrium output Q^*. In other words the individual firm's contribution to the market share is so small that changes in its output would hardly be noticeable. As the firm cannot influence price it must, by definition, be faced with a perfectly elastic demand curve. Moreover, under this model, there is no incentive for the firm to deviate from this market price by overcharging or undercharging for its output. For example, if it accepts the market price of P^* and produces q_1, its total revenue would be equal to the area $0P^*Aq_1$. If, however, it attempted to push prices above the market equilibrium, it would potentially lose all of its customers as they could buy the *same* good or service from one of the many other firms in the industry at the cheaper ruling market price. Moreover, if the firm charged less than P^*, perhaps in the hope of attracting more clients, this would simply lead to a decline in its total revenue as the firm can sell all that it wants to at the market price anyway.

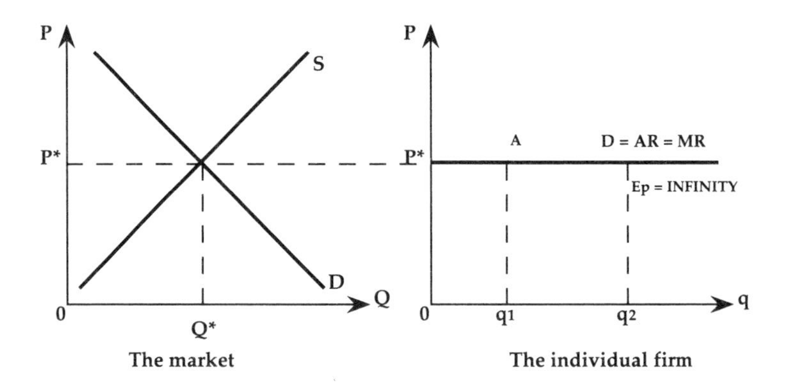

Figure 7.2. *The market and the perfectly competitive firm's demand curve*

Applying this to our example of a firm of house builders, let us take the realistic assumption that there are a number of different house building firms developing sites in a particular town. These developments are likely to contain a highly similar mix of houses so as to conform with planning requirements and current consumer tastes. Thus, the market price for new houses will be set in line with the choice that is available to the consumer, and the available supply from these firms. Moreover, it should be noted that the firms are also in competition with the vast number of 'second hand' houses that will be available on the market through estate agents, or privately arranged sales advertised in newspapers. Thus, if any one building firm tried to increase their prices over the market price, they are likely to lose sales to their competitors as people buy similar houses on a different development. Also, there would be little point in the firm undercutting the 'going' price for a particular type of house if there are a sufficient number of buyers to buy all the houses offered for sale on the market. Such price cutting behaviour as this would only mean a loss in potential revenues. Therefore, it should be apparent that an element of market research is crucial for the building firm to ensure an effective and successful pricing policy. However, the builder would have to recognize that the market price itself is subject to continual change primarily because of changes in the overall level of market demand for such dwellings. Note that with the theoretical extreme of an inelastic demand curve average revenue is constant and the same as marginal revenue as the firm can sell as much as it wants without having to alter the price.

Assumption 2: Product homogeniety

As already suggested, for the theory of perfect competition to work in its purest form, we must also assume that all firms within the industry are producing an **identical product**. This second assumption, combined with the first, ensures that the firm has no control over price. That is, it is also unable to change price by differentiating its product to make it more superior. This assumption may, at first, sound absurd. However, we shall now see how we can apply this assumption to our 'case study' of house builders in a town: one could argue that builders can, and do, differentiate their product. Therefore, although we can look at many house builders producing say three-bedroom semi-detached houses, different building firms may use different designs, different qualities of materials, different standards of finish, landscaping, and so on. In fact, location itself, even within a town, could differentiate one housing estate from another, for example, whether the houses are near an industrial estate, or overlook pleasant views of the countryside or sea, will be key factors in determining the attractiveness of the houses to the potential buyer. Moreover, some building firms may seek to differentiate their product by financial means such as offering special discounts to first-time buyers, or offering subsidized or 'low start' mortgages, for example. Obviously advertising is a key component of such a strategy. However, consumers who are keen just to live in the town, perhaps due to the proximity to their workplace, may be unaware of some of the differences mentioned above. For instance, build quality can be covered up by the finish, and many would not know the difference between a well-built house and one which has not been so well built, simply because few have an intricate knowledge of the building process itself. Furthermore, if building firms are in competition with one another, they are likely to copy each other's 'innovations' so that their developments look equally attractive to clients – if one building firm used extensive landscaping as a positive selling point for their estate, other builders may be forced to follow suit if they found that their market share was declining. Thus, the assumption of product homogeneity may not be as far fetched as originally perceived.

Assumption 3: Perfect ease of entry into and exit from the industry

This assumption means that no barriers to entry or exit exist in the industry. Therefore, if a firm wishes to set up in the industry it can freely do so, just as it could leave the industry if it so desired. This is an important assumption as it ensures that, in the long run, all firms can only make normal profit (see definition in Chapter 5). Normal profit is a level of return which provides just enough reward to keep the entrepreneur in business as it is where total revenue equals total costs, where payments to the entrepreneur are included in total costs. Whilst firms are only making normal profit there is no pressure for firms to enter or leave the industry. However, short-run changes to this situation can be set off due to market led changes creating either short-run profits or short-run losses, and examples of such changes, and their implications, are discussed below.

To understand the importance of this assumption, imagine, at first, that existing firms in the (house building) industry are initially making excess or 'super-normal profits' (super-normal profits are sometimes referred to as 'abnormal profits'). That is, profits in excess of normal profits. Such a situation could have been brought about, for example, by a sudden rise in demand pushing up revenues, or by cost cutting technology that lowers the cost of production. The existence of super-normal profits is shown in Fig. 7.3. Here, each individual firm is receiving the market price of P^* and producing q^* output. Because of their cost structure relative to their revenues, they are making super-normal profits equal to the area CP^*AB. If super-normal profits could be earned in the industry, many would be attracted to join that industry, or existing building firms may direct more resources into the house building sector, so as to take advantage of the high level of profits. As more and more firms enter the industry, or direct resources into it, total industry supply (the market supply of new

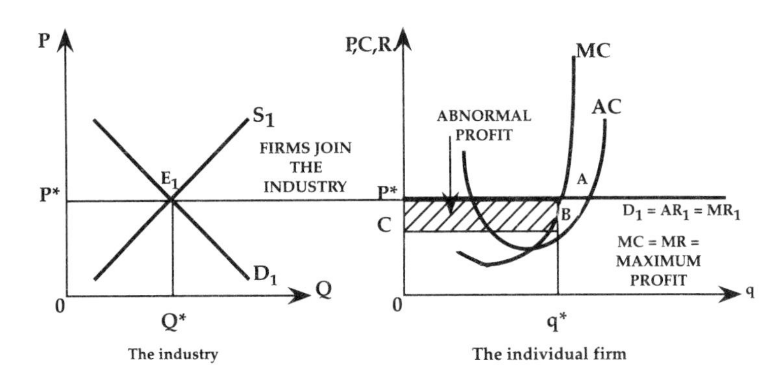

Figure 7.3. *Perfect competition and initial abnormal profits*

dwellings in the town) would begin to shift to the right. In fact, this process would continue as long as super-normal profits could be made and planning permission was granted for such additional developments. However, as the industry supply increases, as more firms join the industry or divert more resources to it, the levels of these additional profits will be driven down as market price is driven down reflecting the increased competition and subsequent consumer choice. The market will then reach the stage where only normal profits will be earned by firms, and thus no further firms will be attracted into the industry. Such a sequence of events can be seen in Fig. 7.4. Figure 7.4 shows that total industry supply has increased from S_1 to S_2 leading to an increase in the industry's overall output from Q^* to Q_2. As more is now available to choose from, prices have been driven down from the original price of P^* to the new, lower, price of P_2. Note also that due to increased competition amongst the firms in the industry, that each firm's individual market share has been reduced from q^* to q_2. With respect to our example of the house builders, high profits being earned in one area are likely to attract more firms to that area. Thus, subject to the availability of building land and appropriate planning permission, the process explained above will be initiated.

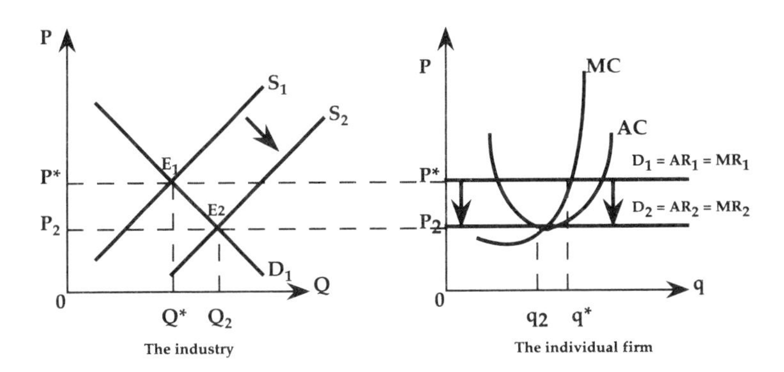

Figure 7.4. *Perfect competition and the reattainment of normal profits*

On the other hand, so as to understand the sequence of events that are likely to occur if a **loss** was being made by the industry, let us now imagine that the house builders in the town, who are presently earning normal profits, are faced with the onset of an economic recession where the demand for housing falls. Figure 7.5 shows that as the demand for housing falls, theory predicts that losses will begin to be made by firms in the industry. That is, building costs remain the same, yet the demand for houses falls from D_1 to D_2, thereby lowering the market price to P_3. Under such circumstances, the output of individual firms drops from q_2 to q_3, and they incur losses equal to the area P_3CAB. In such a situation, some firms could not sustain such losses in the long run and they are likely to go out of

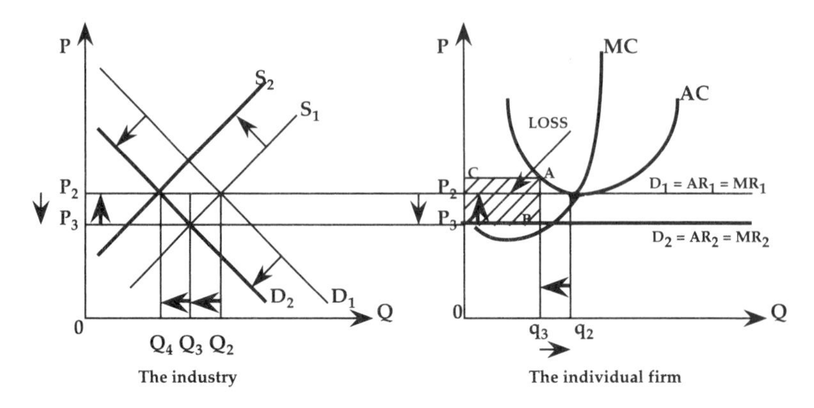

Figure 7.5. *Perfectly competitive firms making a loss*

business. As firms fail and leave the industry, industry supply will be driven back from S_1 to S_2. Such a contraction in supply will reduce output, and therefore choice, which will put upwards pressure on price until we are again in a situation whereby the remaining firms are making normal profit. The end result is that the price of P_2 is re-established although the industry has become leaner as fewer firms means that industry output falls to a level of only Q_4. Therefore, declining market demand is likely to lower house prices and the weaker building firms will go out of business. However, as more firms go out of business, new houses can become short in supply thus driving prices higher again as people who have to move compete for the limited number of dwellings. Therefore, in the long run we are back in the situation where all (surviving) firms are making normal profit. However, in reality, building firms could at least attempt to prevent themselves from going bankrupt in the following ways:

1 Attempt to reduce their cost structure. By looking at Fig. 7.5, one can see that if costs were lowered, losses could be reduced or eradicated. Firms could seek to lower costs by laying off non-essential staff, enhancing productivity with their existing labour and capital, reducing site wastage and theft, using cheaper materials, constructing simpler designs, and so on.

2 Attempt to increase the level of demand facing the firm via the methods of potential differentiation mentioned in the critique of Assumption 2 of the model (see above).

3 Attempt to increase the level of demand facing the industry as a whole. For example, house building firms putting joint pressure on the government to lower the interest rate, thus lowering the cost of borrowing for the purposes of mortgages for house purchase. In fact, the lowering of interest rates will have the dual impact of not only stimulating consumer demand, but also lowering development costs for the building firm as

the cost of borrowed funds in general declines.

Assumption 4: Perfect knowledge on behalf of both producers and consumers

For a truly competitive environment to exist we need to assume that both producers and consumers have a very good (perfect) knowledge of all conditions in the market. In the case of **producers**, as already suggested above in part of the discussion in Assumption 2, the model assumes that all producers have a perfect knowledge of existing and new techniques of production. For example, if one firm were to have a technological breakthrough, all other firms can copy it. That is, there are no 'trade secrets' amongst firms in the industry. With our example of the house building firm, information about such innovations could be gathered through trade magazines, the observation of competitors' sites, plus information passed on by sub-contractors working for more than one firm. Therefore, because no technologically induced cost advantages can be gained by one firm over another, all producers should become equally efficient.

We also need to assume that **consumers** have a perfect knowledge of ruling market prices. Such information would ensure that consumers would not purchase a good or service for a price in excess of the market price as they would know that they could obtain it at a more reasonable price elsewhere. With respect to house purchases, this knowledge could be gained by the potential purchaser through visiting local estate agents, seeking the advice of a surveyor, or reading the property section of the relevant local newspapers. Moreover, as housing is normally an expensive commodity in relation to income, it is highly likely that people will take extreme care in the purchasing process. Due to this fact, people are likely to visit many houses before making a final decision. Thus, consumers should obtain a relatively good knowledge of local property prices. In any case, when an application to borrow money for a house purchase is made, the lending institution would formally send out a surveyor to realistically value the property given current market conditions. In this way an accurate professional assessment of the ruling market price should be obtained independently of any assessment of price by the consumer. This should enable the house purchaser to offer the correct 'market price' for the house. In the case of older properties, the consumer could also instruct a surveyor to undertake a full structural survey of the building. If such a structural survey pointed towards the need for money to be spent on remedial work in the house, such as the need to combat damp, or replace roof tiles, for example, the purchaser could attempt to reduce his offer below the market price so as to make an allowance for these expenses that will have to be incurred at a future date.

Assumption 5: Identical factor prices

Having identical factor prices implies that all firms are placed with the same cost conditions. That is, the cost of raw materials, labour, plant hire, etc. are the same for all firms in the industry. This is quite likely in the house building market, at least at the local level, assuming (as the model does) that all firms are roughly the same size. However, if there are larger firms in the market they are likely to benefit from cost advantages brought about by economies of scale. Such a distortion of the model caused by the existence of a large firm is not that important because there are normally a very large number of firms in competition, and the majority of these firms will be facing similar cost conditions.

Assumption 6: A large number of consumers (purchasers)

For this theory to work in total, we also need to have a situation where there are so many buyers in the market that no one buyer buys a significant proportion of output. Therefore, if one consumer altered their demand, for the good or service in question, it would have such a small impact upon total demand that market prices would not be affected. This condition is relatively easy to uphold in the case of the housing market as few will 'bulk buy' houses! However, this is not to say that such distortions do not exist. House prices have been driven up, for example, if a large firm moves to a town and decides to accommodate its personnel in 'company houses'. In such a case one company may be the sole purchaser of a large amount of housing stock.

Conclusions on the model of perfect competition

In reality, it is unlikely that all of the conditions of the model of perfect competition will hold and as such one will perhaps never observe a 'perfectly competitive' market. However, there are examples of situations where firms operate in such a high degree of competition that this model can be useful to demonstrate the likely behaviour of such firms and their markets. Moreover, the model is used as a 'yardstick' to enable us to see the potential advantages and disadvantages of a perfectly competitive industry. Thus, a government who wishes to encourage increased competition in the economy, for example, could examine the model to see the likely implications of such a policy. Throughout this debate, we have considered the situation of many small firms of house builders competing for their market share in a town. It has also been noted that the analysis may become distorted, although not necessarily nullified, if a large national firm of house builders becomes involved in the market. However, the nearer we do get to this model, the easier it becomes to perceive a situation whereby we will have a large number of similar houses being built by different firms, selling at a price that is primarily determined by the degree of competition amongst firms, and the level of consumer demand.

(B) The theory of monopoly and its application

A monopoly is the extreme situation whereby *the firm is the industry*. That is, there is only one firm operating in that industry. Therefore, the market demand curve will be the firm's demand curve, and for the usual reasons we would expect such a demand curve to be downward sloping (see Chapter 1). The cost and revenue structure of a monopoly is likely to be like that shown in Fig. 7.6. In Fig. 7.6 the firm is producing an output of *Q*, and selling it at a price of *P*. As with the theory of perfect competition, the 'profit maximizing rule' is that the firm will maximize its profits where its marginal cost curve cuts its marginal revenue curve from below as can be seen at point *A*. As with this example, where price (average revenue) exceeds average costs at this profit maximizing level of output, we normally assume that monopolies make super-normal (abnormal) profits. Such profits are over and above the profit necessary to keep the entrepreneur in the industry. However, the existence of such super-normal profits is not always guaranteed for firms in a monopoly situation as this will depend upon the level of demand for their good or service, and their cost structure. For example, if operating costs were to rise significantly, say due to rapidly rising energy costs, a firm could face a loss making situation until it managed to successfully react to the problem, or go out of business. A similar situation could be brought about by a drop in demand. For instance a fall in demand could occur if some consumers choose to boycott the firm's products if they believed them to be 'environmentally unfriendly', or perceived that the firm had operations in a politically unpopular country. At the extreme this could mean that the firm were to

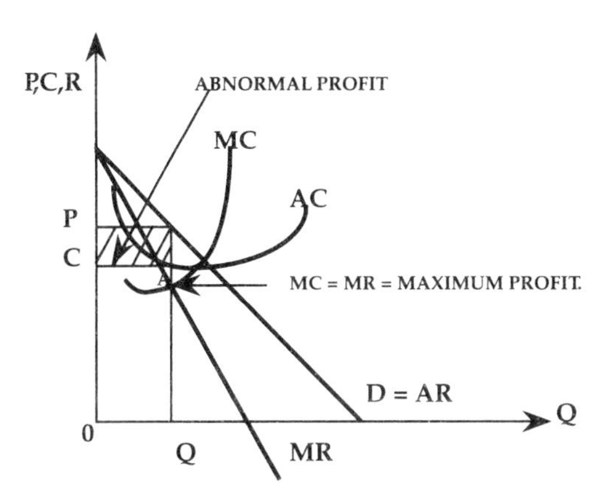

Figure 7.6. *The cost and revenue structure of a monopoly making abnormal profit*

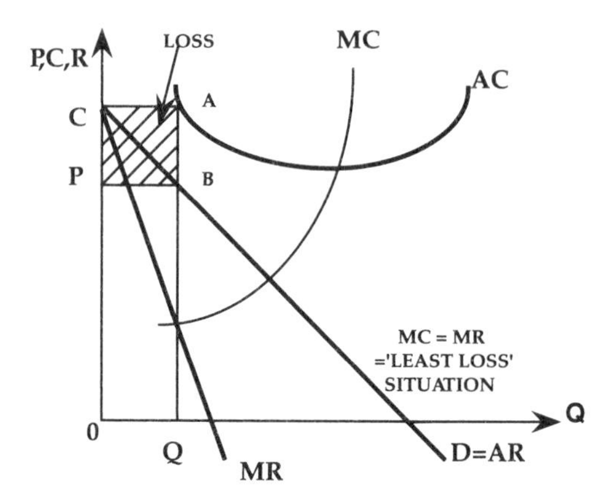

Figure 7.7. *Monopoly: the situation of a monopoly making a loss*

suffer losses. Such a loss making situation is seen in Fig. 7.7. Here, the average costs of production, *C*, are higher than the selling price *P*, of the firm's output, *Q*. As such a loss is being made equal to the area *PCAB*.

In fact, in some countries, government owned monopolies may actually make long-run losses as other policy objectives, other than profit maximization, may be deemed to be more important by the public sector. Such long-run losses can be financed from taxation revenues, or from cross-subsidization from another, profit making, part of the public sector. Alternative goals that the government may wish to achieve are varied. For example, they may wish to keep a loss making industry open so as to secure employment in a depressed region of the country, or they may wish to ensure that the country can produce a good of 'strategic importance' so that they do not have to rely on imports at a time of crisis. The manufacture of defence equipment is a case of this. Also, one can imagine the possible situation whereby a country is unable to acquire cheap imports of a key building material such as cement. Under such circumstances the government may conclude that producing domestically at a loss is the only way of obtaining such an important product in order to facilitate adequate levels of construction in the economy. The domestic industry is likely to incur losses in the first few years of production as the cost of setting up, and providing the necessary capital equipment would have to be covered. Alternatively, even if profits were being made, the government may wish that more is produced and at a lower price. Thus, the firm would be ordered to produce beyond the profit maximizing level as shown in Fig. 7.6, by producing a level of output in excess of *Q* at a price less than *P*. Such action may be felt to be necessary to ensure that an essential service such as public transport is provided in all areas at an affordable price to all members of the community. Whether we examine the scenario of the monopoly making a loss, or suffering reduced profits due to direct government intervention in relation to pricing and output, the industry is likely, at least in the former case, to come under public ownership. That is, such industries would not attract the private monopolist as profits could not be made, and thus to ensure the on-going existence of the industry it would need to be 'nationalized'.

However, if the monopolist does make super-normal profit, in the absence of anti-monopoly legislation or control, he can maintain such profits even in the long run, (Fig. 7.6). This long-run retention of high profits is possible as the existing monopolist can exclude entry into the market by potential competitors because of the existence of **barriers to entry**. Remember that this is in contrast with the theory of perfect competition where only normal profits can be made due to the freedom of entry into and exit from the industry. In the case of a monopoly, many barriers to entry can exist some of which are:

1 The firm may have a unique patent or licence to produce, or provide, a particular good, or service, that it has invented and developed.

2 The monopoly may control all of the raw materials necessary for production via its ownership of them.

3 Specific economies of scale could dictate that there is room for only one efficient firm in the area (in the case of a local monopoly), or in the country (as in the case of a broader, national monopoly).

4 The monopoly may be created by, or at least protected by, the government. This may be done for a variety of reasons as discussed above. A further example could be the preventing of the emergence of a whole range of electricity firms in an area. The rationale for stopping competition in this case would be the fear that having a number of firms may require buildings to be wired up to several different firms so as to give the present, or future, occupier the choice of service. Therefore, government action preventing competition may avoid the wasteful duplication of scarce resources.

Such barriers to entry can give the firm significant **monopoly power**. This monopoly power can be further enhanced by the extent to which the monopoly can control people's purchasing decisions. This latter point primarily depends upon the availability of substitutes for the firm's good or service. For example, a monopoly in wooden window frames is unlikely to be a very powerful one, as many close substitutes to wooden window frames – plastic and metal – do exist. However, a monopoly in the provision of scaffolding, or cement, may be more serious. Thus, monopoly power not only depends upon the control of supply, but also upon the elasticity of demand, where the more inelastic the demand curve facing the firm, the more powerful it can be. Monopoly power is an important consideration in the construction industry as it is often found that monopolies, or at least situations close to monopoly, dominate the materials supply side of the building industry in many countries. Such a market structure can occur in the materials supply industry for a number of reasons:

1 Many building materials and components, such as cement for example, are standard and subsequently relatively homogeneous. In such a situation

there is considerable scope for economies of scale to be had by one large firm.

2 Many construction materials, such as aggregates for example, are heavy and bulky in relation to their value. This feature, coupled with high transport costs, rationalizes the case for local monopolies.

3 Some firms have managed to protect their monopoly position by creating a high degree of 'brand loyalty' for their product. Such brand loyalty may be achieved in a variety of ways, such as through prompt deliveries and back up services, the maintenance and standardization of quality, the provision of short term credit, and so on.

4 Some suppliers may also be the owners of the particular resource. For example a monopoly timber merchant may also, perhaps via vertical integration and the acquisition of another firm, be the owners of the forests from which the wood comes from in the first instance.

5 Cyclical fluctuations in the prosperity of the construction industry inevitably have a 'knock on' effect on the materials supply industry. Thus the materials supply industry must be structured in such a way as to accommodate variable demand. Whereas construction firms normally have the advantage of a low capital base (due to the well-established hire sector that exists for capital equipment for construction in most countries), and a flexible employment structure (due to the hiring of sub-contracted labour rather than full time employees), this is not the case in the materials supply industry. Here, many of the industries require a large amount of plant and machinery and such capital intensity means that the firms are not as well prepared as building firms to cope with short-run changes in output. In order to survive under such circumstances the situation encourages larger firms, thus strengthening the trend towards concentration in the industry.

Before discussing the implications of such monopolies on the construction industry, it is important to note that, in everyday speech, the term 'monopoly' is frequently used rather loosely inasmuch as people are often really talking about situations where a few firms dominate the industry rather than just the one. As such, an investigation into the market structure of 'oligopoly', or other forms of imperfect competition, would be a useful exercise for you to undertake. Having said that, monopolies, and 'near monopolies' do exist, and we will now briefly examine some of the relative disadvantages and advantages of such a restricted market structure.

(1) Arguments against monopolies

When one hears talk about monopolies in general conversation most seem to feel that such firms are an 'evil institution' working against the consumer's interests, and it is in this light that we will firstly examine monopolies. Whilst reading through the arguments put forward below, remember that it is the materials supplier who is often close to being a monopoly, and it is the construction industry who is the 'consumer' in this case. However, it is also important to note that some very large construction firms themselves may be virtual monopolies in specialized markets such as off-shore work, tunnelling, road building, and other key areas of civil engineering. The following is a short list of some of the main arguments against monopolies:

1 Due to their control over the market, monopolies can reduce output and thus force price upwards as the good becomes shorter in supply than would be the case if there were to be higher levels of competition in the industry. Therefore, it is argued that less is produced than would be the case if the industry was in perfect competition, and the higher will be the price. Note that as the demand curve is determined by the consumers, a monopolist can change *either* price *or* quantity and *not* both. In Fig. 7.6, for example, it can be seen that if the firm limits output to *Q* it can charge the high price of *P*. However, if it produced anything in excess of *Q* it would have to lower prices as otherwise it would accumulate unwanted stock. Thus, as can be seen in Fig. 7.8, a monopoly will tend to produce at a high price (*Pm*), and a low quantity (*Qm*) as this reflects the profit maximizing point where its marginal cost curve cuts its marginal revenue curve from below. However, the perfectly competitive firm will tend to produce more (*Qp*) at a lower price (*Pp*). This is because although the perfectly competitive firm will also wish to maximize its profits, and that this is also where its marginal cost curve cuts its marginal revenue curve from below (because its marginal revenue curve is also its demand curve (average revenue) (see back to Fig. 7.3) this will be done where the marginal cost curve cuts the demand curve at a later stage. That is, for the perfectly competitive firm profit maximization is achieved by the meeting of marginal cost and marginal revenue at point *A* in Fig. 7.8 rather than at point *B* as with the monopoly firm. Therefore, with monopoly situations

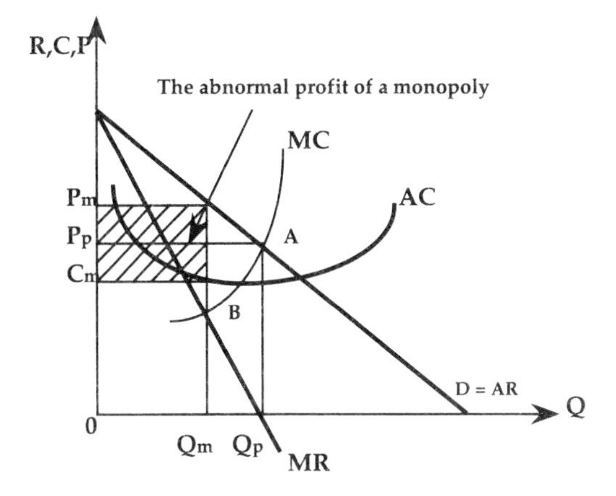

Figure 7.8. *Monopoly and perfect competition compared*

dominating the materials supply industry, building firms may be faced with a poor choice of components at an inflated price.

2 If barriers to entry exist the monopolist may have less incentive to improve the quality of the product as the firm knows that customers are unable to purchase the good or service elsewhere. The degree to which the monopolist can rely on this point will depend upon the degree of monopoly power that the firm enjoys. Moreover, if this argument holds, the logical extension of this analysis is that the monopolist could also become increasingly complacent regarding the quality of its output, the punctuality of its delivery times, and so on. As such, far from the product improving over time, it could become increasingly defective. Therefore, construction firms may have to put up with defective materials that are not delivered on time or in the correct quantities for example. Moreover, they may be unsure of the quality of different consignments of supplies. Such difficulties could add to the time of construction and push up the builder's costs yet further. If such cost increases cannot be passed on to the consumer – due to depressed demand for example – the profitability and survival of some construction firms could be at risk.

3 Continued super-normal profits will generate the monopolist vast resources of financial capital which it could accumulate and then use to increase its monopoly power yet further. This could be achieved, for example, by **horizontal integration** (the buying up of competitors in the same line of business), via **vertical integration** (the buying up of suppliers and/or sales outlets for the product), or by **conglomerate merger** (the buying up of firms not related to the monopoly's original product base). Such merger activity may compound the problems facing building firms suggested in point (2) directly above.

For these reasons, governments have shown concern about monopolies in the private sector, and have often legislated to prevent them being created in the first place, or at least to control the growth and power of those that already exist. In order to achieve this a government body, say a *monopolies and mergers commission*, may be set up to monitor the degree of competition between firms in an industry and its likely impact upon the consumer. However, the power of a monopoly can be eroded *automatically* over time by such factors as:

1 Import penetration. If domestic monopolies push up their prices too far, or let quality slip, it will increase the attractiveness and feasibility of firms using foreign suppliers. However, the degree of such import penetration may be limited due to the low cost, high weight and volume nature of many building supplies such as bricks and cement.

2 The availability of near substitutes. As more alternatives come on to the market consumers would no longer need to rely exclusively on the monopoly's output. For example, a firm that enjoyed a monopoly in the supply of wooden window frames may have to become more competitive as more modern plastic windows come on to the market.

3 The power of the consumers themselves is also an important consideration. The consumers of the monopoly's output could also be large firms and could thus be a major client of the monopoly. Thus, if the monopoly failed to provide such consumers with an adequate service at an acceptable price, this could impair the performance of the construction firm which will in turn damage the materials supplier. Furthermore, it would encourage firms to seek alternative materials or methods of construction so as to remove the reliance on the original monopoly.

Moreover, it should also be noted that there are also some strong arguments in favour of monopolies. A selection of such positive arguments are discussed below.

(2) Arguments in favour of monopolies

Just as there are arguments against this form of market structure, there are also positive sides of the argument. The issues discussed below are perhaps the most supportive views and have been backed up by international research. Your task is to 'weigh up' the arguments for and against this market structure, so as to decide whether restrictive organizations are less, or more, beneficial than a more competitive state.

1 Many argue that the 'lack of incentive' argument (above) is an unfounded one. Rather, it is suggested that large firms are actually more likely to be able to finance expensive research via their ability to fund such activity through the accumulation of super-normal profits. Such research expenditure could have two major benefits for the end user of the firm's product or products:
(i) The existing product or product range of the monopoly could be updated and improved. This should ensure that the construction industry can obtain modern, state of the art materials and that their incorporation could be used as a selling point for their completed buildings. For example, the developer could claim that their office block contains the most modern and efficient services, or it has a very high energy saving rating due to a new, improved form of insulating material used in the buildings construction.
(ii) Technological improvements could alternatively make the monoply's output cheaper to produce. If this were to be the case, the supply curve should shift to the right (see Chapter 1) thus lowering prices to the consumer.

2 As suggested above it would be unwise for a monopoly supplier to behave in such a manner that could jeopardize the very existence of the firms that it supplies. Obviously the survival of both the supplier and the end user are highly interrelated. For example, if a materials supplier were to increase prices too much many construction firms would go out of business. If this were to happen the level of new build is likely to drop and demand would have to be accommodated via more extensive use of

existing buildings, change of use, or refurbishment. None of these alternatives would provide the monopoly with the same level of demand for its output as would be the case for new build.

3 The accusation that a monopoly could reduce the quality of its output may also be unsound. Government regulations could easily be set, and regularly updated if required, to ensure that construction materials were of an adequate, and perhaps improving, standard.

4 The majority of the views against monopolies rest upon the assumption that such firms are making super-normal profits. However, empirical studies have revealed that many such firms actually have sales revenue maximization as their goal rather than profit maximization. The explanations behind this view are varied, but one key theme of empirical studies of such firms is that as management is normally separate to ownership, management will not be obsessed with maximizing profit. Instead it is felt that managers will simply produce so that they are ensured sufficient profits for the survival of the firm and for the satisfaction of its shareholders. Alternative views to explain such behaviour are that firms wish to saturate the market for reasons ranging from longer term profit maximization to mere personal ego. If, in most cases, we do observe sales revenue maximization, rather than profit maximization, it could mean that monopolies may well produce more, and at a lower price, than the initial theory would predict. This debate is summarized in Fig. 7.9.

Figure 7.9 shows that if a firm were to maximize its profits it would produce *Qm* output at a price of *Pm* as this is where its marginal costs were equal to its marginal revenue, or, in other words, where there is the maximum distance between total costs and total revenues. The area of profit is shown in a variety of ways by the shaded parts of the figure. Perhaps the easiest way to see the potential level of profits is by looking at the 'profit bubble' which is simply the area between total costs and total revenues brought down to the horizontal axis. Maximum profits of Π_M would be achieved by the firm producing a level of output equal to *Qm* at a price of *Pm*. However, if shareholders were satisfied by profits of only Πs, the firm could produce more, in this case *Qs*, and at a lower price of *Ps*. Thus, the level of output that the firm produces, and the price that it charges, may in fact be closer to that seen in a competitive market.

Conclusion on the theory of monopoly

Just as with the model of perfect competition, it must be emphasized that the model of monopoly is a theoretical extreme that one is unlikely to often observe in its purest form in the real world. However, many firms do approach such a market structure, and as such this model can show us what that market will be like. The nearer a firm is to being a monopoly the closer the model will be. Moreover, the model helps government decide

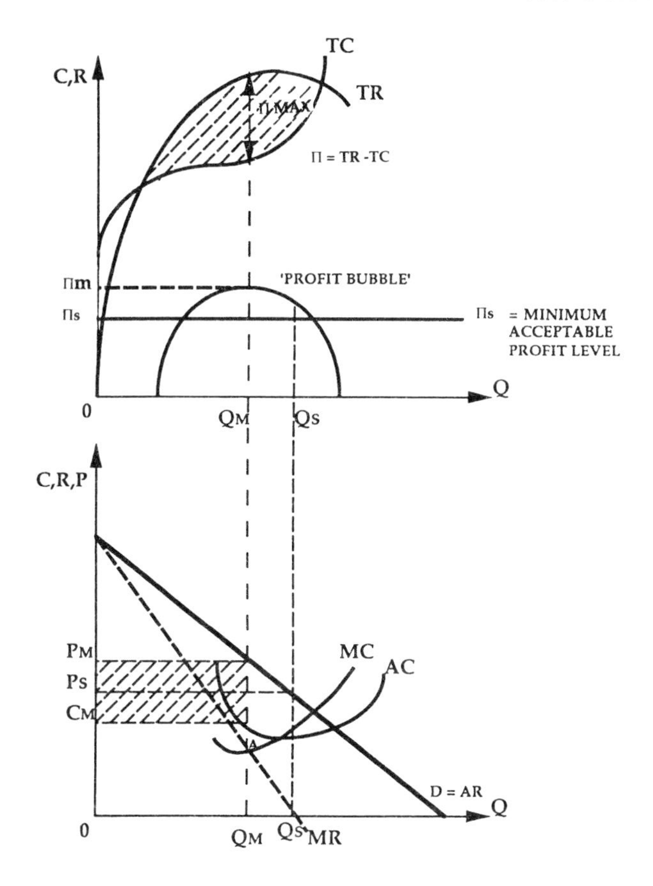

Figure 7.9. *Sales revenue maximization versus profit maximization*

upon whether to encourage competition or restrictive practices. The full impact upon the construction industry of having monopolies in the materials supply sector, or indeed the impact upon the consumer of having some parts of the construction industry itself dominated by near monopolies, depends upon the particular firm or firms in question, and the weight of the arguments posed in the analysis above in each instance. Thus, in order to form an educated opinion extensive research would have to be carried out on a case by case basis.

8

The construction industry

To examine the make-up of the construction industry in a detailed and meaningful way one must clearly define the area of study first. In other words one needs to identify what is meant by the term 'the construction industry' in order to include all the relevant parts of it, but to exclude services that, although related to the construction industry, would be more logically associated with another industry. To help us with this task many countries have official classifications of industries such as a 'standard industrial classification' (SIC). However, care must be exercised when making use of such data as classifications differ between countries, and definitions of industries have changed over time within countries themselves. Perhaps a useful guideline is to limit the definition of the construction industry to those firms involved in the actual *assembly* process of new buildings, and the repair and maintenance of existing buildings. This would then leave out related, but dissimilar industries such as the materials supply industry and the provision of services such as surveying, and development finance.

(A) The construction industry and the economy:

The exact nature of the construction industry will differ from country to country. However, international comparisons do reveal a certain degree of similarity thus enabling one to make the following generalized comments in most cases. It would be a useful exercise for you to try to find data to test the following statements in relation to your country's construction industry.

1 Due to the nature of its product, the construction industry is normally very large in most countries and thus accounts for a high proportion of overall national output. If one adds in the related industries of the materials supply sector, estate agents, and surveying, for example, the significance of construction in the economy is yet further emphasized. One should not be surprised at the magnitude of the industry as one must appreciate that most forms of economic activity require some form of building whether it be an agricultural warehouse, an office, a factory or a retail centre, moreover, everyone needs a home to live in, and there will also be a substantial building stock in existence that requires repair and maintenance, or replacement due to obsolescence.

2 An increasing trend has been the 'globalization' of the construction industry, and its related industries. It has been observed that there has

been an increase in firms competing for projects abroad as well as on the home market. As such many countries have a plethora of multinational construction firms involved in the development process. Similarly, there has been a marked growth in the merging of firms on an international basis as one firm from one country acquires another firm in another country. The importance of such activity is that the construction industry becomes an increasing player in a nation's balance of trade. Essentially, contracts won abroad will bring in export earnings to a country, yet the importation of foreign building materials and services will be a leakage from the domestic economy.

3 The construction industry tends to be highly labour intensive and thus a major source of employment in the economy. Many aspects of construction, such as bricklaying and plumbing for example, would be difficult or impossible to mechanize, or at least it would not be cost effective to do so in most circumstances, and therefore this high dependence upon labour is likely to continue. This is not the case in the building materials supply industry as cement works and brick works for example can be readily automated. Similarly, even the advent of laser measuring equipment in the field of surveying can potentially reduce the demand for trained surveyors as site and building measurement can be done more easily and quickly.

4 In a mixed economy the construction industry plays a crucial role in achieving the government's social objectives. The government requires construction to fulfil its public sector housing programmes, to build hospitals, educational establishments, and provide public buildings, such as libraries for example. However, in many countries there has been a noticeable decline in the public sector's involvement in the provision of goods and services, and therefore the volume of work from the state sector for the construction industry has been declining.

5 Apart from a small degree of prefabrication the product is generally not transportable. This is a feature that is perhaps unique to construction as the majority of other goods can be manufactured in one location and sold in another, and therefore market disequilibrium can be rapidly resolved. This aspect of fixed location in construction means that shortages of certain types of building can exist in one area, yet an excess of floor space can exist in another.

(B) The structure of the private sector of the construction industry

The following points are again generalizations and are not country specific. However, they broadly represent a summary of commonly observed features of the private sector of the construction industry.

1 The private sector of most construction industries are dominated by predominantly small firms.

2 Although building demand in total is geographically dispersed, small, local contracts are the most common source of work for the firm in the private sector.

3 The average number of employees in private sector firms is very low with many employing less than five people. This feature not only reflects the size of firm, but also demonstrates the increasing tendency for the employment of 'sub-contractors' rather than direct labour especially in the case of large contracts.

4 Most private firms tend to specialize in a certain area of construction such as plumbing, wiring or roofing, for example. Generalized data reveal that fewer than half of the firms observed classified themselves as 'general builders'.

5 The 'failure rate' of small construction firms is very high. Such a high number of bankruptcies can be explained by a variety of factors but normally demonstrates the close relationship between prosperity in the construction industry and the performance of the economy at large.

6 Despite the dominance, in terms of numbers, of the small firm, it is usual to find that a small number of very large firms are important in terms of overall construction industry output. In financial terms a major project undertaken by a large firm can be 'worth' many of the smaller projects undertaken by smaller firms. Moreover, the productivity of such large firms is far higher than their small counterparts.

This large number of small firms can be primarily explained by the fact that many of the forces that make it advantageous for a firm in manufacturing to develop into a larger scale enterprise, are not present in the construction industry. The difficulty in the construction industry is that there are few cost advantages of expansion due to the lack of readily accessible economies of scale:

1 Production is based on individual sites rather than in a central manufacturing location. Thus, building firms cannot take advantage of the mass production techniques of most industries.

2 There is a lack of standardization of the product. Each building is different reflecting differing requirements of clients, different locations and varying site conditions.

3 There is a low use of large scale capital as the conditions of construction are not generally favourable to the use and design of such equipment.

4 The system of 'interim project finance' means that large sums of initial financial capital are not always required to undertake the development process. Therefore, even small firms are in a situation to raise the 'start up' finance required for the initiation of most small scale developments.

5 It is unlikely that construction firms will achieve a monopoly situation by gaining control of materials suppliers as building materials suppliers are themselves in a very strong position.

(C) Productivity in the construction industry

Productivity is essentially the measure of output that is produced by various inputs of capital and labour to the productive process. As with all industries, maintaining a high level of productivity in the construction industry is important in order to assure its long run profitability via client satisfaction and speedier debt repayments, and its international competitiveness. However, when attempting to measure productivity a number of difficulties need to be addressed:

1 There are a large number of statistical sources, some of which use different definitions of productivity. For example, some only measure output per actual employee of the construction firm whereas others look at the output of the whole firm including the contribution made by sub-contractors. The difficulty here is that the former figure will overestimate the contribution to total output of the firm by each employee as it does not include the contribution made by sub-contractors hired by that firm in order to achieve its output.

2 Some data sources do not take into account the erosive power of inflation. Therefore, when output is expressed in terms of monetary profit, or turnover, what appears to be constant growth over time may only be reflecting the declining value of money itself. For example, if a small building firm built two houses one year at a cost of £100,000, and the following year they only built one house, yet construction costs had doubled, it would appear that their productivity, measured in terms of materials used in this instance, had remained the same at £100,000, yet it had in fact halved. Therefore, it is best to use data that are given in 'real terms' as such data have taken inflation into account.

3 Financial figures can also be misleading when comparing statistics between different countries as variations in the exchange rate can radically alter the overall view. For example, if a worker in a country is seen to produce an output of $10,000 per year and then the currency of that country is devalued in relation to other currencies, his productivity will be seen to rise in money terms, say to $12,000, yet his physical output could remain unaltered.

4 Different sets of data may contain different definitions of the construction industry itself. As mentioned above one obviously needs to be clear about the exact item that you are measuring otherwise misleading additions or omissions could occur. This problem is especially noticeable when examining international comparisons of productivity, and looking back at time series data in countries where the definitions of industries have changed.

5 Different figures relating to productivity may occur simply because researchers have used different sample sizes. As such one must always take care that one's study sample is statistically significant. Regional

variations in construction productivity may also have an impact upon the overall figure.

6 Evaluations of productivity in the public sector of the construction industry frequently reveal far lower figures than those for the private sector, thus giving 'ammunition' to those who point to inefficiencies in the public sector. However, much of the difference between the two sectors could simply be explained by the fact that the public sector often has a greater load of repair and maintenance work rather than new build and therefore additions to output appear low.

7 It is felt by many that official estimates of construction productivity may be artificially low as firms fail to reveal the true picture of their total output. The reasoning behind this statement is that because of the large number of very small firms (and the growth of self-employment in the industry) many do not make full and accurate returns to the taxation authorities. For example, some builders are well known for 'moonlighting', in other words they undertake work in exchange for cash payments so that taxes can be avoided, and it is in this manner that much small building work may not appear on official records of construction activity.

8 In the northern hemisphere much building site labour is employed on a seasonal basis as most construction work occurs in the warmer parts of the year. Due to the seasonal nature of such employment, output per worker looks significantly lower than for most other industries who are engaged in production all year round in the shelter of a factory or an office. Adding to this problem is the fact that construction contracts are discontinuous and a firm may not be able to guarantee involvement in a new development as soon as another is completed.

So far this analysis has implied a concentration on the productivity of site labour. However, in assessing the efficiency of any building firm one should also look at the productivity of the other factors of production employed by the firm such as capital and entrepreneurship (management). As touched upon above, the importance of enhancing productivity in the construction industry is paramount for a number of reasons.

1 Higher productivity can lower a firm's costs, as, for example, it requires less labour to complete the same task, one needs to hire capital and subcontractors for less time, and there is only a need to borrow development finance for shorter periods. Higher productivity can also increase the revenues of the building firm as there is evidence to support the view that clients are willing to pay more for a job that is done quickly and efficiently as they can then use the building for raising revenue themselves. As revenues are enhanced and costs decline, the profitability of the building firm is increased.

2 Such improved productivity maintains the competitiveness of firms against both domestic and foreign rivals when tendering for work at both home and abroad.

3 It is argued that improvements in productivity help to keep construction costs down and such savings can be passed on to the client in the form of reasonably priced buildings. If the cost of completed buildings were to increase too rapidly when compared with other inflation in the economy there would be a risk that people would 'make do' with older, existing buildings via refurbishment and repair and maintenance, rather than commissioning new build. A commonly cited reason for not ordering a new building is that clients are worried that development costs will escalate during the actual construction phase making the project far more expensive than originally anticipated in their investment appraisal calculations.

4 Some researchers have suggested that high productivity in the construction industry is imperative for high productivity in other areas of the economy. For example, for a manufacturer, or a retailer, to remain competitive the buildings that they require must be completed efficiently and on time. In fact evidence shows that productivity is one of the best indicators of overall economic success as those countries that have experienced high increases in output per person have experienced the greatest economic growth.

As suggested above there are a variety of reasons why construction productivity is difficult to measure and may seem lower than it actually is. However, despite these 'statistical defences' of the performance of the building industry there are also a number of physical characteristics of the construction process that will impair productive efficiency.

1 There are no long, standardized production runs as contracts are varied. Each and every building is on a unique site with its own special characteristics and therefore the production process has to be reorganized again and again. Moreover, even if all building components are completely factory produced, they still have to be assembled on site. Furthermore, although there is much assembly of prefabricated units, much preparatory groundwork is still required. For example:
(i) The site needs to be cleared before construction can commence. This may necessitate the demolition and removal of an existing structure.
(ii) Work needs to be carried out on the foundations of the building. Such work is complicated by the fact that significant variations in site conditions can occur with differing rock structures, terrain, and the presence of mining works.
(iii) The connection of services such as gas, electricity, sewage disposal and water, needs to be organized.
(iv) Access problems to and from the site need to be dealt with. Frequently changes to the urban road system need to be organized in order to accommodate the new development and its likely use.
(v) Even with the best initial surveys and plans unforeseen construction problems can arise on a case by case basis requiring one off 'on the spot' decisions being made.

Coupled to these difficulties is the fact that the individuality of each building design is largely at odds with the standardized, mass produced, nature of goods and services in modern 'industrial societies'. There is also a spatial separation of sites that causes breaks in the flow of production, which subsequently necessitates that both labour and capital need to be mobile.

2 There is discontinuous demand and labour is not necessarily employed all year. In other words, a construction firm cannot guarantee work on a new site when their current development is complete.

3 The completion of a building is often delayed because of frequent changes in specification by either the client or the design team *after* building work has commenced. Such changes are often very difficult, and time consuming to achieve. Moreover, any major changes may have to be agreed upon by the planning authorities which again adds time to the development process.

4 Pressure groups can also delay construction due to particular objections to developments. For example, 'environmentalists' may object to the nature of a building in a certain area, or those who wish to preserve the country's historical past may object to a building that may encroach on a historic site or remains.

5 The planning process itself is often cited as a slow and lengthy one thus adding time between a project's inception and its completion.

6 The construction of buildings is frequently delayed because of slow decision making, and misunderstandings between the design side, the building firm, sub-contractors, and the client.

7 In the case of complex engineering structures and unique designs – that do not use standard construction techniques or materials – the development process can be delayed as unexpected and unknown difficulties are encountered.

8 The types of building that the public or clients desire often do not lend themselves to rapid construction. There tends to be a preference for buildings, especially in the residential sector, to be built in a 'traditional manner' rather than using more modern construction techniques such as the increased use of prefabricated sections.

9 In times of prosperity in the construction industry, so much building work is being undertaken that the supply of crucial skilled labour can become short. In such situations projects have been delayed because key tradesmen – such as plumbers and electricians – cannot come to the site immediately as they are engaged elsewhere. These delays can have a cumulative 'knock on' effect as other tradesmen – such as plasterers – cannot easily perform their tasks until the electricians and plumbers have completed their work.

10 Some researchers have claimed that a further reason for low construction productivity is the poor labour relations and working conditions found on most building sites. Many sites are characterized by a lack of clean, dry, relaxation areas, which most employees in manufacturing would expect to be provided with.

11 There is also evidence to suggest that poor management is another reason for the slow completion of developments. It is a noticeable fact that some firms manage to organize, initiate, and complete production at a far faster rate than others.

Therefore there are a large number of reasons why construction productivity may be lower than productivity levels found elsewhere in the economy. Moreover, some countries may experience even lower productivity levels than others due to such factors as:

1 Building standards or regulations may be higher, or more detailed, in one country than in another. Higher standards are likely to necessitate more careful and detailed building using better building materials, and are therefore likely to add time to the construction process.

2 Poor weather, for example heavy frost, is a frequently cited excuse for delays in the construction process, so much so that building contracts often have clauses in them to allow for delays caused by unfavourable weather conditions.

3 The planning system in some countries is more efficient and less bureaucratic than in others, and therefore delays at the early stage of the project's life can vary because of differences in the planning process.

4 The demands from the public and clients vary from country to country. Increased use of prefabrication may be quite acceptable in one country, whereas in another country, slower, more traditional construction methods will be insisted upon.

One can now see that there are numerous problems to be overcome with respect to construction productivity in order to ensure that the industry is in a suitable state to meet the increasing demands of the modern economy and international competitiveness. As such there now follow some suggestions on how the development process could become more efficient.

1 In many countries it is common to see contracts negotiated between the construction firm and the client whereby it is agreed that any on-going increases in building costs are largely reimbursed by the client. In this way the building firm will not suffer any erosion of its profits if costs increase during the actual period of construction. However, as costs do tend to increase over time due to general inflationary pressures in the economy, one can argue that if the building firm was forced to sign a 'fixed price contract', whereby increases in costs would not be reimbursed by the client, the building firm would have more of an incentive

to complete the development as quickly as possible before cost increases reduced their profit margin on the project.

2 One could have financial penalty clauses written into the contract so that if the construction process was late at any agreed stages, especially at completion, the building firm would have to pay a 'fine' to the client for failing to reach its target objective. In this way one would expect the building firm to have an incentive to enhance its organizational efficiency. In order to implement such a policy the client should be provided with regular 'progress reports'.

3 One could direct efforts at attempting to improve the performance of sub-contractors and materials suppliers as it is often failure on behalf of these groups that lead to construction delays. Perhaps, as with point (2) directly above, the construction firm could impose 'fines' on these parties if they failed to reach their promised deadlines.

4 As many delays are caused by the client themselves, the client must be encouraged to carefully consider their requirements before the construction process commences so as to limit delays created by subsequent variation orders. Moreover, there is often a need for the client to provide more detailed information about their specific requirements so as to ensure that they are catered for.

(D) Innovation in the construction industry

Because of the apparent low level of productivity in the construction industry many perceive it as an industry that is characterized by 'backwardness' and one that is generally very old fashioned. A backward industry can be defined as one that uses out-of-date working practices and techniques, and one that produces a product in an antiquated manner. If an industry is in this situation it is felt that it will not achieve either full *efficiency* or *profitability*. Therefore, there is pressure for construction firms to modernize their operating practices in order meet the requirements of the modern economy and to remain internationally competitive. However, one must note that such a striving for efficiency and profitability may not be in the complete interests of society. Buildings could perhaps be built at a faster rate, however, they may become more basic, lacking in design intricacy, and become heavily reliant on prefabricated components. Such a situation may lead to the creation of a very 'utilitarian' built environment with many similar buildings of cheap construction; although one could perhaps argue that it would be possible to reserve this form of construction for factories on industrial sites – away from the main public view. This argument specifically holds for industrial buildings which may have a very short economic life (see Chapter 13) and therefore it would be expensively extravagant to build such buildings in anything except the most basic way. Despite these fears, and the increasing movement for many new buildings to be built in a more intricate way so as to mimic the architecture of the past, there are still arguments, as seen directly above, for increasing the level of innovation in construction.

Innovation should provide ideas about how to improve the efficiency of working practices, and the techniques of production so that a modern product is supplied on to the market. It should be appreciated that such techniques can be incorporated into the construction process yet at the same time giving the completed building a traditional appearance. For example, a steel framed structure with prefabricated walls could still have a traditionally built and designed facade, or at least one that gave the appearance of being old fashioned. In fact, in many inner city renewal programmes, older buildings have been demolished except for their frontages. These frontages have been retained and a newer structure has been built on to them at the rear. Therefore one has the dual advantage of a modern purpose-built building that is aesthetically pleasing and conforms with the traditional and historic view of its location. When discussing potential efficiency improvements in an international context one must be careful not to make global recommendations as one must recognize that the types of buildings constructed often reflect the relative costs of both techniques of production and materials used. For example, in a country where both bricks and manual site labour are expensive, there is more of an incentive to use alternative materials such as timber which lends itself more readily to prefabrication and subsequently reduces the need for large numbers of site labourers.

Advances in technology can be categorized under a variety of different classifications:

1 modern materials and components technology;
2 enhancing the use of capital;
3 the use of 'Modern Management' techniques.

In the case of *modern materials and components technology* it is felt that increasing the amount of off-site, factory based, production should enhance productivity. In other words, using more prefabricated components should facilitate a speedier construction time, although one would still need the process of assembly to take place on location. In modern times it can be seen that the building process has increasingly become an assembly operation of previously manufactured components, and that such a growth in component technology is likely to take place as long as components are economically transportable from their place of manufacture to building sites. One should remember that advances in technology can make what seems impossible now quite commonplace in the future. For example, at one time even bricks were manufactured on site.

In terms of the degree of *capital input* into the industry, many argue that a low capital to labour ratio makes the construction industry automatically seem backward as most modern industries are highly capital intensive. The **capital to labour ratio** is simply a measure of how much capital is available to each worker in an industry. For example, a low capital:labour ratio would be given by the figure 1/50. This implies that there is only one unit of capital to every fifty workers. Therefore a high capital to labour ratio of say 1/2 implies that there is a unit of capital available for every two workers in the industry. However improving the capital to labour ratio in the construction industry is problematic and possibly not the logical thing to do for a variety of reasons as suggested below:

1 The construction industry is an assembly industry that assembles components that are manufactured by other industries. It is argued that capital intensity in the form of powered plant lends itself easily to manufacturing, yet not so easily to the assembly process. Moreover, manufacturing is often concerned with a homogeneous good whereas construction is usually involved in highly varied production.

2 There is little potential for the application of a great deal of capital in the repair and maintenance sector of the construction industry. Therefore, any data on capital use in the building industry may appear misleadingly low as repair and maintenance is a very large sector of the overall industry.

3 As implied above, the capital to labour ratio will obviously depend upon the relative costs and availability of the various factors of production such as capital and labour. For example, if site labour is more expensive in one country than in another, the capital to labour ratio is likely to be higher in that country as capital usage becomes relatively cheap.

4 The industry is dominated by a large number of small firms, and such small firms will rarely have sufficient finances available to acquire much capital equipment. Moreover, the size of contract that they are involved in is unlikely to be large enough to warrant high capital usage.

5 The need for the ownership of capital equipment by construction firms is not great in many countries as a well-developed hire sector exists to service their requirements. The hire sector has grown as much plant is highly specialized and subsequently too expensive to be afforded by the small firm. Moreover, as firms will normally be involved in a number of different contracts over time they do not want capital to be tied up in a particular piece of equipment that is site or job specific. Furthermore, construction contracts are often discontinuous, and therefore firms will not wish to incur the expense of having capital equipment lying idle in the event of them not securing a contract immediately after the completion of the existing one. Finally, building firms are reluctant to purchase equipment that is easily damaged in the rough environment of a building site. Therefore, if one examines data on the use of capital in the construction process it may give an artificially low figure as the capital is owned by firms in the hire sector rather than the building firms themselves.

Alternatively, some argue that the way forward to improve productivity in construction is for the industry to take on board *'modern' management technology*. The theory of modern management has its origins in the early part of the twentieth century when there was an increasing need for improvements to be made in manufacturing productivity so as to cater for the mass production necessary to meet the needs of rapidly escalating international demand. The name most frequently associated with these ideas is Frederick Winslow Taylor leading to the concept often given the name 'Taylorism'. The main thrusts of modern management theory are:

1 One should organize the labour force in such a manner that there is a **division of labour** leading to the **specialization of labour**. Therefore, instead of hiring workers who would perform a variety of tasks on site from bricklaying to plumbing, one should have specialists who only performed one task. In this way it is hoped that workers will become highly skilled, and therefore highly efficient, when performing their particular task. The additional motive behind this idea was that it would *deskill* the workforce converting all labour's output into physical, non-mental energy that would not challenge the managerial hierarchy or its authority. The idea of increasing output via specialization was certainly not new as it had already been discussed by Adam Smith in 1776 in relation to his example of a worker in a pin factory in his book *An enquiry into the nature and causes of the Wealth of Nations* (Routledge, 1890). However, the negative aspect of such specialization is that workers can become bored by their repetitive tasks and therefore they may begin to work in a slow, despondent way paying little attention to the pride and quality of their work.

2 Management should undertake 'time and motion studies' in order to determine a fully productive workload for the labour force. In this way management could then attempt to control tasks by the speed of machinery, such as conveyor belts, and the succession of individual components to complete the final good.

However, one could argue that the ideas of modern management are *already* incorporated into construction management. Consider the following examples:

1 There is a clear division of labour into the various parts of the construction process such as plasterers, carpenters, electricians and bricklayers.

2 Wages for work such as bricklaying are often paid on an incentive or bonus basis that encourages workers to complete tasks as quickly as possible.

3 Although methods of production that use conveyor belts are not strictly possible in the construction industry as the product is stationary, one can make the labour force pass along the product line with sequential deadlines to complete their individual tasks. In this way there are great similarities in the productive process of both manufacturing and building except that in the case of construction labour passes the product rather than the other way around.

Such ideas to improve productivity via advances in technology also need to be supplemented by the targeting of other issues that tend to retard productive efficiency in the construction industry. Delays in the building process have also been blamed upon the following influences:

1 The climate is a frequently blamed variable for the slow completion of a building. The changing of seasons can have an impact upon both the climate itself, and the number of hours of daylight. Moreover, there are daily variations in the weather which can hinder production. For example, heavy ground frosts can make any ground work – such as the digging of foundations – virtually impossible. Therefore, unlike the enclosed conditions of the factory environment that one associates with the manufacturing industry, the construction industry is open to the natural elements which affect standardized production and uniform working conditions. Even if advances in building technology can reduce the problems created by the weather, one still has to be concerned with related issues – such as the safety of workers in poor light for example.

2 Architects producing unviable, and over complicated ideas, or not making their designs clear enough at the outset so that remedial work needs to be carried out during the construction process.

3 The planning process is often criticized for being slow and bureaucratic which can lead to delays in commencing construction after the inception of a project.

4 Poor management and the lack of co-ordination of sub-contractors is frequently blamed for the slow completion of buildings.

5 One group of sub-contractors damaging the work of another group of sub-contractors is another frequently cited cause of concern that again leads to the need for remedial work to correct the damage.

6 New materials, or materials that have been used incorrectly, can lead to the threat of failure, and therefore there is a need to correct the work before the completion of the building.

In conclusion on the topic of productivity in the construction industry the following points can be made:

1 Due to economic pressures changes in the method and type of construction are only likely to occur if such actions enhance the profitability of building firms.

2 There is increasing social pressure from the general public, who form the construction industry's client base, and in particular from groups such as the environmentalist movement, to construct new buildings in a 'traditional' manner and design, and to conserve existing buildings to a greater extent. However, the use of such traditional techniques could prevent the incorporation of productivity enhancing techniques in the construction industry.

3 Many feel that innovation is necessary though as building costs often increase more rapidly than the general level of inflation. Therefore, in order to maintain output, builders need to find ways of keeping prices down, via innovation, in order to maintain demand.

4 Many building firms are reluctant to use new technology, or new products, due to the degree of risk involved in using new ideas. This is especially important in the building industry due to the large sums of money involved in construction, and the small financial reserves of most small building firms.

(E) Overseas construction opportunities for domestic firms

Some domestic construction firms may feel that they are sufficiently experienced to tender for projects abroad as well as at home. Foreign contracts will become attractive if the level of profitability from such work seems to be greater than that offered by building for the domestic market. However, it should be noted that there can be many problems with overseas investment especially because such an involvement often carry far greater levels of risk (see below). Despite this there has been a noticeable growth in the number of construction firms operating on an international basis, and this growth can be attributable to a number of factors.

1 During a depressed domestic market, overseas opportunities can look more rewarding, assuming that there is not a complete global economic slump. Different parts of the world experience economic prosperity and recession at various times and at different stages of their development, and as such markets that show great market potential now may not always do so, and there can be some promising markets in the future that are not yet established as an area of great construction activity and growth.

2 Due to the enormous wealth generated in the oil producing areas of the Middle East, overseas construction orders have tended to reflect the state of the oil market. Thus, when the price of oil is high, oil producing nations will receive more income which can then be spent on their economic development.

3 The emergence of 'newly industrialized countries' (NICs), especially in the Far East, has led to the very rapid development of these areas. Obviously a rapidly expanding economy will require a rapidly expanding building stock.

In obtaining overseas contracts the firm must be able to demonstrate that it is in a better situation to undertake the project than a domestic firm in that country. In relation to this point a number of issues can be raised:

1 The size of firm is normally a 'barrier to entry' to overseas work. Research has shown that most work put out to international tender is

concerned with very large projects for which only the largest firms are best suited to compete. Typically, the problems of project size and risk are reduced via joint ventures with other firms. Moreover, more experience of a country can be gained by formally merging with domestic firms or by opening a local subsidiary.

2 Foreign firms are in a better position to obtain contracts when there is a lack of indigenous construction technology and management skills in the country in question. Although smaller projects can be undertaken by local builders, they may lack the necessary expertise for larger work.

3 The 'rules' of fair international competition are sometimes distorted as governments lend assistance to their own nation's construction firms by providing them with a number of advantages in an effort to secure overseas orders. For example:
(i) The provision of subsidies to the construction firm in order to reduce its costs, or to offset the potentially higher level of risk associated with much overseas work.
(ii) Governments have increasingly 'tied' any financial aid that they have given so that the recipient country has to spend the aid that it receives on using firms and products from the donor country.
(iii) Similarly, it has been suggested that 'diplomatic muscle' has been put on some countries in order to pressurize them to award the contract to a specific country.

4 To ensure enhanced profitability, many firms have undertaken the practice of **transfer pricing** which is essentially an accountancy procedure designed to lower the total tax burden of multinational companies (MNCs). Transfer pricing enables intracorporate sales and purchases to be artificially invoiced so that profits accrue to those branch offices located in low tax countries, while offices in high tax countries show little, or no, taxable profits.

However, before becoming involved in overseas work the construction firm must be made aware that a number of serious difficulties often occur with such contracts. These problems are most marked, and frequently encountered, in the lesser developed countries (LDCs) of the world, and thus the existence of these problems in such areas increases the attractiveness of more stable markets in the more developed countries. It should be noted that:

1 International competition is very fierce, with some countries benefiting from cheaper sources of project finance, lower labour costs, and higher levels of construction productivity and so on.

2 Some large construction firms have exposed themselves to serious financial problems after going ahead with projects when they have received insufficient and unreliable local information. For example it is important to get a clear picture on all aspects of the building process ranging from likely increases in building costs to the eventual market for the completed development.

3 Political turbulence in some countries has meant the abandonment or delay of some projects. Moreover, in the case of a change in government, or revolution, there have been examples of new rulers refusing, or failing to meet the commitments of the previous government. In this way previously ordered construction projects may not be paid for, at least on their previously agreed terms.

4 Financial crises have occurred where the overseas client has gone bankrupt and therefore cannot purchase the completed development. Alternatively, if the building firm is working in conjunction with a group of local firms, financial difficulties with any one of these firms can cause the collapse of the whole consortium.

5 High inflation in many countries can erode the value of projects agreed in local terms, and can lead to large increases in building costs. The majority of projects take a number of years to complete and inflation rates in excess of twenty per cent, for example, have not been uncommon in the past.

6 Foreign currencies can also depreciate in relation to other currencies reflecting a poor local economy. Indeed currencies are often deliberately devalued as devaluation is a method of making a country's exports cheaper, (an injection into the economy), and imports more expensive (a leakage from the economy). For example, one may negotiate the contract at an exchange rate of £1 = $1, where the '£' symbol represents the currency of the home country of the construction firm, and the '$' symbol represents the currency of the overseas nation in which the contract is being undertaken. Therefore, if the contract is worth $1m. the firm should receive £1m. when it repatriates its earnings to its home country, (obviously this figure will be affected by taxation in both countries). However, if the '$' is devalued so that we have a new exchange rate of £1 = $2, on conversion of profits at the completion of development the construction firm would now only receive £0.5m. for its $1m.

7 Rigid 'exchange controls' can hinder the free flow of both project finance and the repatriation of profits. Such controls are often highly bureaucratic in nature and can lead to costly project delays, or even the cancellation of potential projects. For example, an application may need to be submitted to the country's central bank so that money can be released to purchase essential building materials or services from abroad. If there is a delay in this procedure it could hold up the whole project because of the sequential nature of construction tasks, otherwise alternative, second best, solutions may have to be tried.

8 Some countries often seek to impose stringent conditions on foreign firms who gain access to their markets. For example, they commonly insist that local contractors and suppliers are used, as well as promoting joint ventures with domestic engineers and surveyors. This condition should not represent a problem in itself, but frequently construction

firms have been let down by underqualified local contacts or constraints in the local supply process such as the adequate delivery of materials.

9 Poor quality of local construction materials has also been a major cause of concern, especially with respect to on-going quality control, and delivery times.

10 National or local laws, customs, and working practices can lead to misunderstandings and project delays.

Despite these potential difficulties it should be noted that they are variable over time and between countries. Moreover, the growth in the 'globalization' of construction work suggests that these problems are not insurmountable and that the benefits frequently outweigh the negative aspects of such work. The advantages to countries of awarding contracts to overseas construction firms is that they can provide both foreign skills and capital, where indigenous firms may not yet be big enough to take on all contracts, especially the larger, more complicated ones. However, there are also many criticisms of multinational firms and their operations. For example, MNCs are unlikely to reinvest profits in the 'host' country as they export them to the parent company. Moreover, elaborate structures built by such firms often necessitates the importation of materials and expatriate skills that can use up a country's valuable foreign exchange. In addition to this, research has revealed that many MNCs do not greatly contribute to tax revenues due to the practice of transfer pricing, and the fact that they are often given attractive tax concessions and subsidies in order to encourage their participation in the project. Moreover, the operation and success of foreign firms could inhibit the growth of domestic firms.

CONCLUSION

This section has been designed to give you an insight into a variety of economic forces that shape the behaviour of firms in the built environment. The analysis has ranged from the examination of the costs and revenues facing the individual firm, to the market in which the firm operates. Importantly, application of the theory presented here enables one to judge how firms will react under a number of circumstances, and how the firms themselves can influence the market. A useful exercise for you to undertake now would be to re-examine the questions at the start of this section to see how economic theory can be used to examine a variety of key applied issues.

PART 3

The macroeconomy and the built environment

The analysis so far in this book can be categorized as **microeconomics** because it is involved in the examination of specific economic aspects in isolation. For example, microeconomics enables us to examine the likely prospects of a small building firm entering a highly competitive industry (see Chapter 7), or, we could look at the market for a particular type of housing, or office accommodation in a certain town, etc. (see Chapter 3). Now that we have gained this knowledge, and understood the methods of analysis involved, we can bring these issues together to look at the economy in its entirety. That is, we move on to the study of **macroeconomics**.

In this part of the book, therefore, we will be looking at a variety of large scale issues and their impact upon the property and construction industries, as well as the built environment in general. Chapter 9 will examine the general workings of the economy, whilst highlighting economic problems that can and do occur. This analysis will be kept to a general level as it is felt that the application of the theories, rather than an intricate knowledge of their detail, is far more useful to the student of real estate issues who is unlikely to want to become an economic theoretician. Chapter 10 suggests likely scenarios for firms directly involved in, or dependent upon, the construction industry, if the economy experiences difficulties such as high rates of inflation or unemployment for example. Finally, Chapter 11 will look into how governments can attempt to manipulate the economy, at both the national and local level, in order to achieve economic and political objectives.

9

The economy: a simple macroeconomic model

There are a wide variety of models available to us to explain how the economy operates. These models range from the very simple to the highly complex. However, the objective of this book is to keep such models as basic as possible so that they can be easily understood and readily applied. Moreover, so as to utilize your knowledge gained so far, theories learnt in the microeconomics section of this book (see Parts 1 and 2) will now be built upon to examine macroeconomic issues and their relationship with the built environment. Although by adopting this more simplistic stance it must be appreciated that some loss of detail and accuracy may occur, the basic principles expounded by the following analysis are sound.

As an initial starting point one must appreciate that in an *open*, and *mixed* (macro)economy, as is the situation in most countries of the world, there are four main components of the economy: consumers, firms, government and foreign trade. An **open economy** is one which trades with other countries, and a **mixed economy** is one where both the public and private sectors are involved in the actual workings of the economy. This chapter will now briefly discuss how these components of the economy interact so as to give us a simple working model of the economy itself. This model can be formulated from straightforward demand and supply analysis. Chapters 10 and 11 will then go on to examine how government policy, at both the national and local levels, can be used to attempt to intervene in, and regulate, the economy.

(A) Demand in the economy

(1) Consumer demand

In any country there will be a large number of people who, as long as they have some form of purchasing power (and do not rely solely on food aid for example), will enter into a large number of daily transactions ranging from the buying of necessities to the purchase of expensive consumer durables. Such demand may depend upon a large number of variables such as incomes, expectations, government policy, and the desire to save, to name just a few.

With respect to incomes, if consumers have a high 'propensity to consume' it means that they spend the large majority of their income and save little. In fact, as we shall see shortly when we introduce the concept of the **multiplier process**, a useful economic concept is the **marginal propensity to consume** (*mpc*). The marginal propensity to consume is simply a measure of how much consumers spend out of any increment in their incomes. For example, an *mpc* of 0.8 suggests that for every extra one per cent increase in income, consumers will spend eighty per cent of each one per cent rise, and thus only save twenty per cent of each one per cent rise. Therefore, as one can either save or consume, the **marginal propensity to save** can be found by subtracting the marginal propensity to consume from one. That is if

$$mpc = 0.8$$
$$mps = 1 - 0.8 = 0.2$$

It should be noted that a consumer's income can come from a variety of sources and not just wages. For example income can be gained from social security payments to the unemployed (where available), interest received from monies held in deposit accounts, dividends from shares, etc.

Looking at the variable of 'expectations', it can be found that it is likely that consumers will delay consumption if they expect prices to fall in the future, or, increase consumption now if they expect prices to rise in the future, or shortages to occur. Unfortunately, both of these forms of expectation led behaviour can be self-fulfilling prophecies. For example, if people increase their spending now, due to a fear of future shortages and rising prices, demand will go up thus putting upwards pressure on prices and existing supplies and in this way are likely to cause shortages which may not have occurred in the first place.

The latter chapters in this part will deal in detail with the influence of government policy. At this stage, however, it should be sufficient to state that a government could attempt to influence consumer demand by altering the ease of the availability of credit, changing the cost of borrowing via changes in the interest rate, or changing the level of taxes for example.

Relating consumption expenditure to the built environment, it can immediately be seen that the higher the level of consumer demand for goods and services, the higher the need for retailing and industrial output, and thus the greater the need for buildings (shops and factories) in which such economic activity is carried out. The existing domestic supply of these buildings may not be able to cope with very high levels of demand, or sudden increases in demand. This supply constraint will come about if there is insufficient capacity in the current buildings to meet the new, higher levels of demand, as it will take time for extensions to existing buildings to be built, or new premises to be found. Moreover, new businesses may well need to be set up to accommodate this demand, and such new enterprises will not materialize 'overnight'. Therefore, in the immediate period we could expect, for example, overcrowded shops before more retail outlets were built to accommodate higher consumer demand. Furthermore, as we could expect little increase in supply from the domestic industrial sector, this could lead to an increased volume of goods from abroad, imported by retailers in an attempt to fill the gap created by excess demand. Such imports could worsen the country's balance of trade, yet the alternative is higher inflation as prices rise in an attempt to ration out current supply amongst the high levels of demand.

Also, as we shall shortly see when we discuss the 'multiplier process', a high level of consumption expenditure is likely to lead to rising incomes. As incomes increase, more people are likely to be able to enter the housing market, afford better, bigger houses, or have extensions built on to their existing houses. Therefore, we could expect increased activity in the residential sector as well as that seen in the industrial and retailing sectors.

In other words, increasing demand tends to lead to an overall 'property boom'. However, it must be noted that the overall impact of rising demand on each part of the property market can be severely distorted by social factors. The following two examples will clarify this statement. Firstly, it has been observed in some countries, that increases in incomes are largely spent on housing rather than in any other sector of the economy. This primarily reflects a desire for owner occupation, and means that manufacturing industry, apart from those industries directly involved with the housing sector – such as furniture and carpeting – will not experience any great increases in demand and corresponding growth. In such circumstances we would expect there to be a boom in both the residential market and some parts of the retailing sector of the property market, but not the in the industrial sector. A second example that could lead to little growth in the industrial market would be that some countries have a very high 'propensity to import'. In such instances people have a preference for imported goods over those produced domestically. The reasons for such preferences can be varied and range from the fact that it is simply fashionable to own imported goods, or that some imported goods are deemed to be better in terms of quality, for example, than similar locally manufactured goods. If such import penetration were to occur, any increases in incomes are likely to be spent on foreign goods and such expenditure will only benefit the industrial sectors in foreign countries rather than the industrial property market of the 'home' country.

Finally, it should be noted that, as well as the above, consumption expenditure can be influenced by long run structural changes in the economy such as changes in the distribution of income, or demographic changes. For example, if there was a significant level of internal migration within a country from a depressed region to a more prosperous one, we would see growth in the property market, in terms of prices and new building, in the region to which people were migrating, whereas there would be a relative 'collapse' in the market in the area which people were leaving. This would have an impact upon all types of property as not only would more housing be needed in the preferred region, but more people moving to that area would also be likely to stimulate retail and industrial growth in the area due to the concentration of their purchasing power. The selling price, or rent levels, of existing buildings in this area would increase reflecting the increased competition for their occupation. However, the depressed region would now become even more of an economic 'backwater'. As people moved away, the demand for houses would drop and house prices would decline. Moreover, a great deal of the area's spending power would be lost to the prosperous region, and as such retailers and industrialists in the depressed region may need to reduce output or even close down.

(2) Demand from firms

Obviously firms themselves need to demand goods and services in order to function as a business. Thus, a large number of the demands in an economy are from this sub-sector. For example, in order to construct buildings, a construction firm will order raw materials from other firms, and may also need to invest in new machinery as existing equipment wears out and needs to be replaced. Note that in both of these instances we are dealing with a 'derived demand' as firms do not order these items as an end product, but require them to produce the buildings for which there is a demand. For example, as demand in the economy grows during a period of economic recovery, we would expect more building work to be ordered (see point 1, Consumer Demand, above) and thus construction firms, and the businesses supplying those firms, should be faced with the good financial prospects of rising orders. This is again a process that will be stimulated by the multiplier effect (see pp. 153–158). However, again it should be noted that, much of this 'inter-firm' demand may 'leak' abroad as construction firms, for example, order materials and components from overseas suppliers. Such a propensity to import by the industry may be based upon arguments of quicker delivery times, better quality, improved choice, and so on.

(3) Demand from government

A large volume of overall demand in most countries comes from the public sector at both the local and national level. The volume of such demand will depend upon a variety of factors, with the two main forces being:

1 The current state of the economy and the government's reaction to it. For example, a country experiencing a deep economic depression may need strong government intervention to put it 'on the road to recovery' (see Chapter 11).

2 The political philosophy of the government in question. Some governments, for example, are very keen to promote largely unrestricted economic activity by the private sector, whereas others believe in a great degree of public involvement in the economy.

However, in most countries, the government sector will be involved in the provision of the main road networks, constructing educational establishments such as schools and universities, building prisons, hospitals, defence installations, and so on. Therefore, construction firms, in particular, will receive a large volume of demand from this sector. Again a multiplier effect will be apparent (see later). As well as reading the later chapter of this part for more information on the impact of government activity in the economy it will also be useful to see the information on 'market failure' and 'public goods' in Chapter 15.

(4) Demand from abroad (exports)

Another potential source of demand in the economy comes from abroad. That is, other countries will demand goods and services from this country. In order to attract, or maintain, such demand, a nation needs to ensure that its goods and services are internationally competitive in terms of both price and quality. This could be achieved by competing on grounds of efficiency, technological lead, quality, etc. Moreover, good management of the exchange rate by government is also required so as to ensure that the country's exports are competitive in terms of price. However, it should also be noted that although a country may win many export orders, much of the demand created by such activity, and its associated positive impact upon the economy, may be reduced or even eliminated by domestic demand 'leaking' abroad to buy foreign goods. Therefore, the net effect on the country's economy will depend upon the overall balance of trade.

In terms of construction. exports would be represented by domestic firms securing contracts overseas. However, the net effect on the economy of such activity by the construction sector would have to be viewed against all the materials and services that the building industry imports from abroad.

A conclusion on demand in the economy

In order to measure the total level of demand in an economy we need to measure all of the sub-sectors discussed above. Once we have achieved this we can attempt to aggregate them together in order to create an **aggregate demand function** for the whole economy. Thus, we could arrive at the expression:

$$AD = C + I + G + (X - M)$$

where AD = aggregate demand;
C = consumption demand;
I = investment demand;
G = government demand;
X = exports;
M = imports.

(B) Supply in the economy

As with all market analysis, we need to consider both sides of the market. That is, we need to look at supply as well as demand in order to execute meaningful analysis. Therefore, we will now go on to briefly examine the sectors of the economy that supply goods and services. Once we have done this it will enable us to derive an **aggregate supply schedule** using the same methodology as we did with our analysis on the demand side of the economy.

(1) Supply from consumers

The assets that an individual can supply on to the market are their labour, or their entrepreneurial skills. Therefore the amount of people willing to work in an economy at a given set of wages, or reward, will have a great influence upon the overall level of supply in that country. For example, in a country affected by lethargy, a poor state of general health, or compulsory military service, a large proportion of the labour force will not be able to provide their productive services, and therefore overall economic output (aggregate supply) will be impaired. Likewise, a country that does not invest in an adequate education system may fail to produce entrepreneurs to initiate, or advance, the supply process. In fact, it may lose entrepreneurs if it fails to give them sufficient financial and/or material incentives to remain in their home country. This problem can sometimes be seen as a 'brain drain' as top professionals leave their country to take advantage of more favourable circumstances in another country, such as lower levels of personal taxation, or the provision of superior research facilities for example. Moreover, without a good education system, the country may not create a sufficient pool of skilled labour that would be required to undertake many forms of production.

(2) Supply from firms

Supply from firms is perhaps the most obvious form of supply as it is the firms that provide the goods and services that we demand as consumers. The success, and therefore the output, of such firms will depend upon a variety of key economic variables ranging from the commercial taxation structure, the state of the economy itself, the cost of finance, to the structure of the industries themselves – whether they are highly competitive in nature or not.

(3) Government supply

As shown in the section on aggregate demand (above), the government is frequently involved in the provision of a large number of goods and services ranging from the operation of nationalized industries, to the provision of public utilities.

(4) Imports

Another source of supply into the economy is the volume of imports from abroad. If a country has a good export performance, it can afford to make up any inadequacies in domestic supply via this route. However, if imports constantly outweigh exports the country may soon suffer from a balance of trade crisis. In the construction sector one can witness imports in the form of imported materials and equipment, but also by foreign firms being involved in development work in this country.

A conclusion on supply

Once we have measured the output from all of the above sectors we can add them together in order to generate an **aggregate supply function**.

(C) A simple demand and supply model of the macroeconomy

Given full information concerning aggregate demand and aggregate supply we can now produce a simple demand and supply model that will:

1 provide us with a framework to examine the economy as a whole,
2 provide us with a means to specifically analyse the impact of government intervention upon the economy.

This model is represented in Fig. 9.1. In Fig. 9.1, the curve *AS* represents the aggregate supply function, and the various curves labelled *AD* represent differing levels of aggregate demand that may be experienced in an

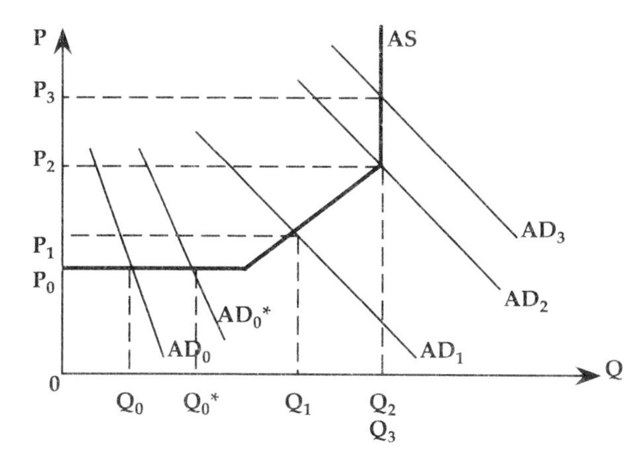

Figure 9.1. *Aggregate demand and aggregate supply: a simple model of the economy*

economy, ranging from the low demand depicted by the curve AD_0 to the high level of demand shown by the curve AD_3. Thus, it can be seen that, given the existing aggregate supply function, changes in aggregate demand will cause changes in the level of output in the economy (Q) and in the general level of prices (P). With the economy represented in this manner the following points should be borne in mind.

Firstly, it can be seen that the aggregate demand function is downwards sloping. This reflects the normal inverse relationship between price and quantity as people economize or postpone consumption if prices are too high (see Chapter 1). Secondly, aggregate demand can shift depending upon the variety of influences that affect consumers, investors, government decision makers and exports as was suggested above. For example, if aggregate demand was depressed during an economic recession, it would be represented by a low aggregate demand function such as AD_0. However, as the economy began to recover, the curve would shift to the right representing higher levels of overall demand. Thirdly, although a nation's ability to produce, and therefore the position of its aggregate supply curve, is relatively fixed in the short run, it can shift. Such movement could be due to government 'supply side' policies (see Chapter 11), technological changes, environmental concerns about current techniques of production, and so on. Fourthly, it can be seen that the aggregate supply function is characterized by three distinct sections. These sections range from an initial perfectly elastic portion at low levels of output, to a perfectly inelastic portion at higher levels of output.

Thus, the shape of the aggregate supply function suggests that at very low levels of output, any increases in demand can be accommodated by utilizing excess capacity and therefore there will be no inflationary pressure. This can be seen in Fig. 9.1 where aggregate demand increases from AD_0 to $AD_0{}^*$ leading to a rise in national output from Q_0 to $Q_0{}^*$, but prices remain unchanged at P_0. In terms of the built environment, one can explain this by pointing out that during a recession, or period of low demand, it is highly likely that there will be vacant or under utilized industrial and commercial properties. Therefore, as aggregate demand increases, and more industrial and commercial space is required, firms will be able to utilize the existing accommodation, and will not be competing for a limited amount of space which would drive the sale price and rents of such properties upwards. It should be noted, however, that this broad discussion relates to the economy in general, and therefore may disguise the fact that some regions of a country may suffer from a lack of appropriate buildings, or may suffer less from a recession and its correspondingly depressed demand. In such regions, any growth in aggregate demand will lead to competition for space that will have an inflationary impact with buildings being sold or rented to the highest bidder. These cases are either so few, or they do not deviate greatly from the norm, that they do not greatly influence our overall picture of the economy. In fact, if one were to have a region of the country whose performance differed to a *very* large extent from the performance of the rest of the economy, it could be argued that it should be excluded from the general analysis, and be treated as a 'special case', otherwise it could distort the overall analysis.

Returning though to the general picture of the economy shown in Fig. 9.1, one can see that as demand begins to rise further, say from $AD_0{}^*$ to AD_1, shortages of supply will begin to threaten as the aggregate supply function becomes relatively inelastic. Once the economy has reached this stage the available supply will be rationed out by the price mechanism leading to an increase in the general level of prices from P_0 to P_1. In fact, if aggregate demand were to rise too rapidly, or too far, and we reached a level of aggregate demand shown by AD_2, or above, the economy would experience inflation with no corresponding increases in output. Note that as demand increases from AD_2 to AD_3, output remains constant at Q_3, yet prices rise from P_2 to P_3. Such inflationary pressures could be reduced if action was taken to boost aggregate supply by building more factories, increasing productivity, and so on.

Therefore, it can be seen that neither an extreme of low, or high, demand is good for the economy as a whole. At very low levels of demand, such as AD_0, we would experience little, or no, inflation, but a high under-utilization, and thus under employment, of resources would exist. In such a situation we could expect to see evidence of depression such as derelict factories and vacant office buildings, and people out of work. Whereas at very high levels of demand, such as AD_2 and AD_3, we could have the full employment of factors of production, such as capital and labour, given the current state of technology, and accepted procedures, yet there would be high inflation such as that shown by the price levels P_2 and P_3.

As changes in aggregate demand and aggregate supply can have such far reaching consequences on the economy, most governments seek to exert some form of control on these two forces. To enable governments to attempt such intervention they have at their disposal both **demand management** policies and **supply side economics**. Before examining such government manipulation of the economy, and its implications to the property and construction industries, a good understanding of the 'multiplier process' is required.

The multiplier process

The 'multiplier', as the name suggests, is a process that magnifies the net impact of an alteration in the economy. The multiplier effect can work in both an *upwards*, positive manner, and a *downwards*, negative manner. For example, if we increase an 'injection' into the economy, such as an increase in government expenditure, the multiplier process will be positive and increase incomes by a greater amount than the initial value of the injection. Conversely, if we increase a 'leakage' or outflow from the economy, such as increased expenditure on importing foreign goods, the multiplier will be negative and will tend to create a downwardly spiralling effect of declining incomes. To understand the mechanisms involved in this process we shall now examine applied examples for both the positive and negative multiplier.

(1) The positive (upwards) multiplier

Positive multipliers are induced when there is an initial *injection* of money into the economy. Such injections can be created by an increase in government expenditure, an increase in the level of investment, or an increase in the volume of export orders. For example, if the government were to increase its expenditure on the repair and maintenance of public buildings, the mechanisms behind the multiplier process would suggest that the overall impact on the economy would be far greater than the initial rise in government expenditure. The reasoning behind this result can be demonstrated by going through the following sequence of events: if the government spends more money on public buildings, such as schools and libraries for example, construction firms would have to be employed to carry out the work. These construction firms would have to order materials from firms in the materials supply industry who may need to take on more workers in order to cope with the increased orders. Moreover, the building firms themselves may need to take on more workers so that they can undertake the contracts. Therefore, we now have more people employed in both the construction industry and the materials supply industry. Many of those who now have a job in one of these industries may have been previously unemployed, so that now they are earning an income. Thus, with more people in employment we would expect more to be spent in the shops as more goods and services are purchased. The retail sector may then need to order more supplies from industry as stocks are depleted, and they may also need to take on more staff to accommodate the increased sales volume. These new staff employed in the retail sector will spend their money in the shops, and another similar sequence of events is initiated. Thus, it can be seen that consumer prosperity can lead to prosperity in both the retail property market and the industrial property market. Therefore, an injection of money into the economy could lead to a far greater level of net income due to the series of 'chain reactions' that such an injection will set off.

Because of this central role of consumer expenditure in the multiplier process, when quantifying the magnitude of the multiplier one needs to find information on the nation's marginal propensity to consume (*mpc*), as well as the actual value of the initial injection. Once one has obtained these data, the following simple formula can be used to ascertain the actual value of the multiplier:

$$K_u = \frac{d.\,Injection}{(1 - mpc)}$$

where K = the symbol for the multiplier;
u = the notation to indicate that the multiplier is 'upwards';
d = change in;
mpc = the marginal propensity to consume.

For example, assuming an injection of £100 million of government expenditure, and that the marginal propensity to consume in the economy is equal to 0.6, we could insert these figures into the equation in the following way:

$$Ku = \frac{d.\,G}{(1 - mpc)} = \frac{\text{£}100\text{m}}{(1 - 0.6)} = \underline{+\text{£}250\text{m}}$$

Therefore, although government expenditure was only increased by £100 million, because of the chain reaction that is set off, the overall level of activity in the economy has been increased by a further £150 million so as to produce a total effect of an additional £250 million in the economy. Thus, as the value of the initial injection has been multiplied by two and a half times, in this example, the value of the multiplier is 2.5. Alternatively one could find the value of the positive multiplier in the following way and then multiply the answer by the level of the injection:

$$Ku = \frac{1}{(1 - mpc)} = \underline{2.5}$$

A knowledge of the value of the multiplier is important for the government in terms of promoting growth and employment, but also when trying to control inflationary pressures in the economy. For those involved in the property world, an idea of the magnitude of the multiplier would be useful when trying to determine the level of new development that may be required from the stimulation in activity caused by any increases in injections that occur in the economy. However, it must be appreciated that the full impact of any positive multipliers may be counteracted by 'leakages' from the economy creating 'negative' or 'downwards' multipliers.

(2) The negative (downwards) multiplier

Negative multipliers are induced when there is a withdrawal, or leakage, whereby spending power is taken out of the home economy. Examples of such leakages are:

1 People's spending power being reduced by taxation. As money is taken away from consumers in the form of taxes, like income tax, and even local property taxes, they have less money to spend and thus consumer demand falls.

2 Expenditure on imported goods and services means that consumer demand does not benefit local businesses and therefore the positive impact of such expenditure will be felt abroad rather than at home.

3 If people were to save a higher proportion of their income than they do at present less money would be available for expenditure and therefore demand would fall. (Although, in the very long run, demand may increase because if savings are wisely invested they would create increased purchasing power for the saver in the future.)

The mechanism behind this negative multiplier is very similar to that of the positive multiplier, except that it occurs in the reverse direction. This can be seen in the following way: as a leakage increases and consumer demand falls, less will be spent by consumers on goods and services. Therefore, for example, the retail sector will record a drop in sales, and offices will conduct less business. If this occurs these businesses may need to lay off workers who are no longer needed, and order less from their suppliers in the manufacturing sector. As the demand for industrial output declines, firms in this sector may also need to make some staff redundant. As more people become unemployed, they too spend less in the shops, and as such the downwards spiral continues. For example, if taxes are raised, perhaps to finance increased government expenditure, people's disposable income will be reduced and thus so will their spending power. Therefore, a negative multiplier will be induced. In an attempt to quantify the net downwards effect caused by this mechanism, one again needs to gather information on the value of the current marginal propensity to consume, and the actual level of the initial withdrawal, (which is a tax rise in this example). Once one has these data it can be put into the following formula which would show the overall magnitude, and net effect of such a leakage:

$$Kd = \frac{-mpc.\ d.leakage}{(1 - mpc)}$$

Therefore, for example, if the overall tax burden on the economy had risen by £80 million, and the marginal propensity to consume was equal to 0.6, one could insert these values into this formula in order to estimate how much of a deflationary effect this would have on the economy:

$$Kd = \frac{-mpc.\ d\ T}{(1 - mpc)}$$

$$Kd = \frac{-0.6\ .\ £80\text{m}}{(1 - 0.6)} = \underline{-£120\text{m}}$$

Therefore, because of the sequence of events that are set off by the negative multiplier (described directly above), the leakage has produced pressures to decrease overall income by £120 million rather than just the £80 million of the tax. In other words, the net effect is 1.5 times greater than the actual leakage. Thus, in this example, the value of the negative multiplier is 1.5. Alternatively, we could have found the value of the negative multiplier in the following way, and then multiplied the answer by the level of the initial withdrawal:

$$Kd = \frac{-mpc.\ 1}{(1 - mpc)}$$

Thus, in our example:

$$Kd = \frac{-0.6.\ 1}{(1 - 0.6)} = -1.5$$

Likewise, if £20 million had been spent on imports, the formula would suggest that because of the sequence of events that are 'triggered off' by such leakages, the overall negative effect on the economy would be far more damaging than this initial figure would indicate, as domestic industries may close down for example. Again one could calculate the full impact of such expenditures by inserting the relevant values into the formula:

$$Kd = \frac{-\,mpc \,.\, dM}{(1 - mpc)}$$

where: M = imports.

$$Kd = \frac{-\,0.6 \,.\, £20\text{m}}{(1 - 0.6)} = \underline{-\,£30\text{m}}$$

Therefore, direct leakages will create a negative impact on the overall income in the economy. Such a downturn in activity could also be caused by an increase in the level of savings, as any increase in the marginal propensity to save will cause a reduction in the marginal propensity to consume. In other words, less money will be available to consumers to spend upon present consumption. Thus, in our example, if the marginal propensity to save increases from 0.4 to 0.5, it would imply that people now saved fifty per cent of their incomes rather than forty per cent, and therefore the marginal propensity to consume would fall from 0.6 to 0.5. If this were to occur it would reduce the strength of the upwards multiplier as can be seen from the following numerical examples.

If government expenditure was increased by £100 million and the marginal propensity to consume was 0.6, as we have already seen, the net effect on the economy was to increase incomes by £250 million, or by two and a half times:

$$Ku = \frac{£100\text{m}}{(1 - 0.6)} = \underline{+£250\text{m}}$$

However, if the marginal propensity to save were increased to 0.5, the marginal propensity to consume would drop to 0.5 so that the increase in government expenditure would only increase by a magnitude of two rather than two and a half:

$$Ku = \frac{£100\text{m}}{(1 - 0.5)} = \underline{+£200\text{m}}$$

Therefore, an increase in the marginal propensity to save means that less money is being spent in the shops, retailers are ordering less from industry, and so on. Note also that such a decline in the marginal propensity to consume caused by higher savings would also reduce the magnitude of the negative impact of the downwards multipliers. This can be explained by the fact that less would now be spent upon such leakages as imports, and less would be received in the form of sales tax.

A conclusion on the multiplier effect

It is highly unlikely that one would observe an injection into the economy, or a leakage from it, and their associated multipliers, in isolation. In reality leakages from an economy are often offset by corresponding injections. For example high taxation gives a government more revenue in order to increase government spending, high levels of savings means that financial institutions have more money to lend out for the purposes of investment, and by receiving monies from exports a country is more able to pay for imports. Therefore, one needs to examine the full range of alterations in the economy so as to appreciate the net effect. For purposes of illustration, imagine that the government had increased its expenditure by £100 million, and in order to help finance this they have raised taxation by £80 million. Moreover, assume that £20 million of this rise in government expenditure is to be used for the purchase of foreign goods. The net impact on the economy could then be assessed using our formulae:

$$\text{Net impact} = \frac{dG}{(1 - mpc)} + \frac{-mpc.dT}{(1 - mpc)} + \frac{-mpc.dM}{(1 - mpc)}$$

$$\text{Net impact} = \frac{£100\text{m}}{(1 - 0.6)} + \frac{-0.6.£80\text{m}}{(1 - 0.6)} + \frac{-0.6.£20\text{m}}{(1 - 0.6)}$$

$$\text{Net impact} = £250\text{m} - £120\text{m} - £30\text{m} = \underline{£100\text{m}}$$

Given a knowledge of these forces the professional in the property market should be able to gauge the likely impact on the economy, and thus the property market, of any changes such as a rise in imports, an increase in taxes, a fall in government expenditure, etc.

10

Primary economic objectives and the construction industry

In order to induce a strong economic climate, and to maximize the well-being of those who live and work in the economy, it is felt that a number of key macroeconomic objectives need to be achieved. These objectives are:

1 a steady rate of economic growth;
2 a low, and stable rate of inflation;
3 a low level of unemployment;
4 a balance of trade.

This part of the section will now briefly examine each of these objectives in turn, especially in relation to their importance to the construction industry.

(A) A steady rate of economic growth

Economic growth essentially means that the level of activity in an economy increases over time as more transactions take place and more is produced. Such growth will cause an increase in the demand for goods and services in general, and more people are likely to be employed in an attempt to ensure that supply meets the new, higher levels of demand. As more people are employed, they now receive an income, and therefore their expenditure is also likely to rise, and we can see that the multiplier process has begun to operate. This positive growth is beneficial to the construction industry in a number of ways. Firstly, increased activity in the commercial sector is likely to create increased orders for office, retail, and manufacturing buildings, although the level of these orders would depend upon the level of excess capacity that exists in the economy, and how much demand 'leaks' abroad to foreign suppliers. Secondly, a healthier economy will give rise to higher incomes which could stimulate the residential new build market as more people can afford a home of their own, or move into a larger house. Moreover, higher incomes enable the owners of buildings to spend more upon the repair and maintenance of their existing building stock.

If a steady state of economic growth is not achieved, and either an economic decline or stagnation is observed, it is obvious that the construction industry will suffer a reduction in orders as it will largely be catering only for the essential replacement of buildings. Such a state of economic decline could have long run effects on the built environment as property developers and industrialists lose their confidence in the economy to achieve a favourable investment climate. Moreover, if an economy actually experiences negative rates of economic 'growth', many firms will either go out of business or will be operating below their full potential. In such circumstances one would experience an excess supply of buildings in the commercial sector and at the extreme derelict buildings would be observed.

It is important to note that although economic growth is, of course, desirable, it must be a *steady* rate of growth that is not too slow, too fast or too variable. If the level of growth in an economy were too slow it is unlikely to attract foreign investment as investors see the likelihood of more promising and higher returns elsewhere. On the other hand, if the rate of growth was very high it is likely to lead to inflationary pressures in the economy which in themselves could be detrimental to the long-run prosperity of that economy. Moreover, fluctuating rates of growth are also likely to deter the investor even though the long-run trend may be encouraging. Investors need to be cautious in an environment where large sums of money are being risked, and the timing of the returns so crucial. Economic growth can also lead to the inequalities within an economy increasing as some people, or some regions of a country, benefit from such growth to a greater degree than others.

Achieving a steady state of growth is far from easy, especially when governments also have to be aware of the distributional impacts of such a policy. Moreover, carefully laid out plans for growth have frequently been nullified by the emergence of unanticipated 'exogenous shocks' to the economy for example – such as increases in the cost of energy. Furthermore, policy makers must ensure that neither the poor implementation of policy, or the failure of policy itself, actually retards the rate of growth that would have occurred in the absence of such public intervention.

(B) A low, and steady, rate of inflation

Inflation is essentially the process by which prices rise in an economy and is normally measured by a 'retail price index' (RPI) expressed as an annual percentage. The retail price index is normally a measure of how prices change from month to month for a typical weighted 'basket of goods' that are bought by the average family. The figures raised by such a measure should be treated with great caution and must only be used as a guideline. The reasoning behind this statement is that the figure will only be 100 per cent accurate if you are the average householder purchasing the same bundle of goods as selected by the statisticians calculating the RPI. Obviously, great variations do occur, and more detailed measures of regional rates of inflation, or rates affecting different income groups, or different industries, may be more appropriate. A useful, and interesting, task is for you to try to calculate the rate of inflation that affects you and compare this with the official retail price index figures. International comparisons should also be treated with caution as different countries will select different items than others in order to calculate their inflation rate and place different weightings on the importance of various items. Indeed, even looking at inflation over time within a country can be misleading as the make up of the representative bundle of goods can also change over time. It is usually possible to obtain separate information on construction related rates of inflation such as building cost increases, house price changes, and rental growth, for instance.

The goal of low and steady inflation is again largely designed to produce confidence in the economy. For example, if such a goal is achieved an investor should be able to assess with a degree of confidence the likely value of a future stream of returns from an investment after the erosive power of inflation has been taken into account. High, or volatile, rates of inflation will obviously make such calculations more difficult and subsequently investment becomes increasingly risky and less attractive. A stable rate of inflation is perhaps more beneficial to an economy than having no inflation at all. The rationale behind this argument is that if prices consistently rise by say five per cent, and if producers cannot pass all of this increase on to the consumer, they will have to devise more efficient means of production in order to maintain their profitability levels in the light of rising prices.

If a low level of inflation is not achieved and price rises *exceed* increases in wages, people will experience a decline in purchasing power which will result in reduced levels of demand for goods and services in the economy. Due to the workings of the negative multiplier effect this will obviously have a detrimental impact upon the construction industry as fewer new buildings are required due to the reduction in economic activity. Inflation can also lead to a redistribution of income: typically the better off in society are in a stronger position to defend themselves against inflation and therefore one may witness the building of luxury homes and prestige offices, yet at the lower end of the income scale, few will now be able to afford their basic 'starter home'. The direct impact of inflation upon building firms is that construction cost increases push up costs which are unlikely to be mirrored by increases in revenues as demand is most probably dropping rather than rising. Such a reduction in profitability can mean that marginal projects are not undertaken, and marginal firms go out of business. However, if a firm has already commenced a project and high inflation occurs during construction, it is probably best to complete the development and sell it in order to recoup some costs, rather than being left with a half completed site and no revenue.

(C) A low level of unemployment

'Unemployment' is essentially a measure of the number of people who are officially out of work. As with inflation this figure can be misleading as the definitions of unemployment are different between countries, and have been changed over time within countries themselves. Moreover, many who may not wish to register as being unemployed will not appear in the official statistics, and as such government data may be an understatement of the actual reality. If there are a large number of people unemployed, they will be on low incomes and therefore there will be a lower level of economic activity than would have been the case if all were in productive employment. This reduction in national income will initiate a negative multiplier process that will have a detrimental impact upon the construction industry. Moreover, unemployment is frequently a regional feature with higher rates experienced in 'depressed regions' than those seen in more 'prosperous regions'. Therefore, it is not uncommon to witness large volumes of construction activity in one area, yet blight and dereliction in another. Generally though, high levels of unemployment lead to a reduction in incomes and a subsequent decline in construction orders, and because of the labour intensive nature of the construction process, the building industry typically suffers unemployment rates two to three times higher than the economy on average.

(D) A balance of trade

A balance of trade occurs whereby a country manages to pay for all of its imports from its earnings from exports. Temporary surpluses or deficits are not seen as a major problem; however if either condition is allowed to persist difficulties may arise. If a long run **deficit** were to occur leakages from the economy are greater than corresponding injections as imports are greater than exports. As consumers spend more upon foreign goods, the demand for locally produced output falls and therefore a decline in the domestic manufacturing sector is inevitable. Initially this will lead to a reduction in construction orders in the manufacturing sector, however declining orders could be experienced in all sectors as a negative multiplier process is initiated. Eventually such a deficit would have to be corrected by government if the country were not to fall further into debt. On the other hand, a persistent **surplus** on the balance of trade account can also be detrimental to the economy's performance in the long run. Seeing an economy in surplus is likely to stimulate a favourable image of strength in the economy in the eyes of the external investor and therefore attract an inflow of investment funds into the country. This inflow of 'hot money' into the economy is likely to increase the value of the domestic currency in relation to other currencies as people invest in it. As the value of the domestic currency increases the construction industry can be affected in two ways: Firstly, *imported* construction materials and services now become relatively cheap as more can now be purchased from abroad due to the relative strength of the domestic currency. In such situations, therefore one should expect an increasing number of foreign items in domestic buildings as overseas products become more price competitive with local products. For example, imagine that the local currency is denoted by the symbol '£' and that the currency of a leading trading partner is denoted by the symbol '$'.

If originally the exchange rate was £1 = $1 one would need £100,000 to purchase $100,000 of construction materials from abroad. However, if the value of the domestic currency were to rise so that the new exchange rate were £1 = $2 one could now purchase the $100,000 worth of construction materials for only £50,000 thus making the imported goods far cheaper than before. The problem with this is that domestic building materials suppliers will lose orders and this is likely to induce a negative multiplier effect. The second impact upon the construction industry is that a rising exchange rate will reduce its *export* potential. For example, at the initial exchange rate of parity, an overseas project yielding a profit of $1 million will produce a profit of £1 million at home. However, as the value of the domestic currency rises, the value of the overseas profits will drop when they are repatriated to the home country as $1 million will only be worth £500,000. Therefore, under such circumstances many overseas projects will become unattractive investment proposals.

One can now see that a variety of important economic objectives need to be striven for in order to achieve a healthy economy. Moreover, the health of the construction industry, and those industries related to it, whether one looks at chartered surveying or the building materials supply industry as examples, are inextricably dependent upon the health of the economy itself.

— 11 —

National and local government economic policy and the built environment

Just by reading the newspapers and by watching the television news, one should be aware that all economies suffer from periods of recession, when growth is low and unemployment is high, followed by periods of recovery and relative economic prosperity, which can lead to escalating inflation, and rising imports, if the economy begins to 'overheat'. This process tends to repeat itself over the years so that the economy proceeds, usually with some underlying growth, in a cyclical manner. This cyclical path of economic activity, and the terminology used to describe its different stages, is shown in Fig. 11.1. It can be seen that when an economy is experiencing a **recession** the level of economic activity is falling, whereas a **depression** is the very depths of the recession. When the economy experiences more rapid growth after a depression it is said to be in a **recovery**, where the very peak of such a **recovery** is known as an economic **boom**. The average rate of **growth** that this economy achieves is given by the growth path **gg** in Fig. 11.1. As already implied, it should be noted that either extreme of the cycle, 'trough' or 'peak', may be disadvantageous to the health of the economy. Moreover, a rapidly fluctuating economy will lose the confidence of investors who will be too unsure of the future to commit themselves in any great way. Thus, in order to attempt to achieve a stable and steady rate

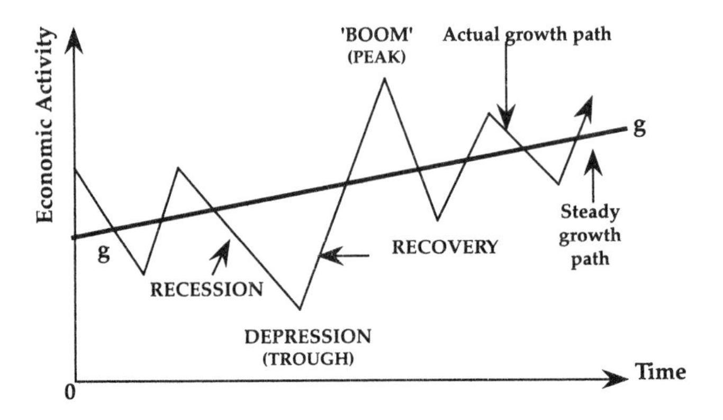

Figure 11.1. *The economy's cyclical growth path*

of growth, the government has a variety of economic management tools at its disposal to attempt to reduce both the frequency and amplitude of such cycles. By managing the economy the government would hope to promote a strong, confident economic environment.

Such intervention in the macroeconomy by the government can be via the use of 'demand management' policies, and/or 'supply side' policies. Both of these techniques of cyclical management are discussed in detail below. In this analysis it should become apparent that the construction industry itself can be targeted and manipulated by government economic policy. Moreover, because of the size of the construction industry, and the nature of its product, inasmuch as most economic activity takes place in buildings, it is easily effected by such policies.

It should be noted that the impact and effectiveness of such economic policy is often reduced due to the fact that much government activity is of a *political nature* as well as an *economic* one. In fact some argue that most government actions are solely politically motivated and can cause long-run damage to the economy in exchange for brief, 'vote catching' short-run economic successes.

(A) Demand management

Demand management is where the government tries to manipulate the level of **aggregate demand** in the economy. The two main policies that are available for such demand management are **fiscal policy** and **monetary policy**. These policies can be used independently, but it is more likely that they will be used in conjunction with one another in order to achieve overall economic objectives. The emphasis on fiscal policy or monetary policy will depend upon the ideology of the government using the policy, as well as the nature of the task that demand management is being used for. The reasoning behind this statement should become apparent in the following analysis as fiscal policy and monetary policy are examined in detail. To fully understand how these policies can be used one needs to refer to Fig. 9.1 to see movements in the aggregate demand schedule as we follow through an economic cycle in Fig. 11.1 from a depression to a boom to a recession.

(1) Fiscal policy

Fiscal policy is a demand management technique whereby governments *directly intervene* in the running of the economy by using the **instruments** of changing **government expenditure** and/or the level of **taxation**. Because of this direct approach, fiscal policy is seen as being a more *interventionist policy* than monetary policy. It is for this reason that a reliance on fiscal policy is often associated with governments that are to the 'left of centre' in political terms. Although, it should be noted that most governments, whether on the 'left' or 'right' of the political spectrum will use a degree of fiscal policy in their management of the economy. Fiscal adjustments to the level of government expenditure and tax levels are often made in annual **budgets**, or announcements by the minister of finance, or equivalent (chancellor), that are designed to **fine tune** the economy. Therefore, in summary, fiscal policy can be seen to have two instruments:

$$Fiscal\ Policy = dG\ and/or\ dT$$

where d = change in;
G = government expenditure;
T = taxation.

We shall now examine how fiscal policy can be used during an economic cycle, and see how such policy affects the construction industry and the property market.

(a) Recession

If the economy is experiencing a recession, or worse a depression, aggregate demand will be low and could be represented by the aggregate demand curve AD_0 in Fig. 9.1. This situation of low aggregate demand is normally referred to as a **deflationary gap**. That is, there is a gap between where aggregate demand should be for a healthy economy, and where it actually is. Such depressed demand could have been induced by an escalation of world energy costs, for example, leaving consumers and firms little spending power or investment finance. As the level of demand drops in the economy fewer transactions take place in the retailing, office and industrial sectors, which is likely to lead to people becoming unemployed and some firms going out of business. Therefore, in terms of the built environment we are unlikely to see many new buildings being constructed, conversely the slowdown in business is likely to lead to vacant or even derelict commercial buildings, and a fall in real rental values as demand declines. In the residential market, house prices are likely to drop as people's incomes decrease, and those who have been made redundant, and who did not take out mortgage insurance, may be unable to keep paying off their mortgage and thus end up having their home repossessed. Moreover, in very severe recessions, many may be unable to afford even basic repair and maintenance expenditure on their buildings which could lead to an unsafe and unsightly building stock. If the economy reaches such a situation, the nation is under-utilizing its resources, and is suffering from the 'social cost' of unemployment, and the 'eyesores' and related social problems of unkept or unused buildings, to name just a few problems. These problems tend to lead to further difficulties such as an increased crime rate and other more serious urban disorder. Therefore, governments could attempt to improve the situation by initiating an **expansionary fiscal policy** aimed at increasing the level of economic activity and overall prosperity. These policies are hoped to have an **inflationary effect** on the level of aggregate demand. Such an expansionary demand management policy could be achieved by increasing government expenditure and/or decreasing taxation. Both of these instruments will now be examined in order to assess their effectiveness in promoting the desired economic growth:

(i) Increasing government expenditure

Generally, if a government increases its expenditure it will normally lead to more public sector projects being undertaken. If the public sector has to employ more people, or contract the work out to private firms, in order to achieve such projects, both the people and the firms will experience rising incomes or profits. Therefore, one would expect the positive multiplier process to be initiated. For example, a building firm contracted to build a public hospital will take on more labourers, or employ more sub-contractors, and order the necessary materials and plant. Thus, unemployed workers may now be put in productive jobs, they will spend more in the shops, the shops will need to order more stock from industry, and so on. Likewise, the materials supply sector may have to increase production, especially if the building of this hospital in our example is just a small part of overall increases in government expenditure on a larger building programme. Therefore, the materials supply industries will take on more workers, and the various stages of the positive multiplier process will again be set off.

To analyse the implications of increased government expenditure in more depth it is important to distinguish between two categories of public sector spending. Firstly, there is **capital expenditure**. Capital expenditure is money spent on new construction work such as roads, schools and hospitals (as per the example directly above). Moreover, one also finds under this classification capital grants for buildings in the private sector. Such grants may be issued to private firms to help them with, or totally cover, the cost of setting up a factory, for example, in an economically depressed part of the country. However, it should be noted that not all capital expenditure is money spent directly on the built environment. Data for capital expenditure can often include government spending on new machinery, and vehicles, etc. Despite this though, it is clear that increased capital expenditure by government will lead to more work for construction firms in the public sector. The second type of government spending is **current expenditure**. Current expenditure is money spent on the running of government, the civil service, and any nationalized industries. Thus, rather than being individual payments for 'one off' projects – of which construction is one example – as is the case with capital expenditure, current expenditure is concerned with meeting ongoing costs. However, it still is an important source of potential work for the construction industry, as capital expenditure can include for example:

1 Money for housing subsidies and improvement grants. These are often made available to people who own homes that do not conform to current, acceptable standards. For example homes that suffer from damp, structural problems, or lack basic amenities such as an indoor bathroom, may be eligible for such a grant.

2 Money for the maintenance of local authority buildings such as hospitals and schools.

3 All construction for the military is included as current expenditure in some countries.

In fact, even the on-going payment of public sector employees, as part of current expenditure, can have a positive, yet indirect, effect upon the level of building work. For example, if civil servants receive a large pay award they are likely to spend more in the shops, and the positive multiplier process will again be initiated.

(ii) Decreasing taxation

Just as with increasing government expenditure, the lowering of taxation is an attempt to stimulate higher levels of aggregate demand. Here we must remember that taxation is levied upon both consumers (income tax, sales tax, etc.), and firms (corporation tax, etc.). With respect to consumers, if taxes are reduced they will have more money left over (disposable income) to spend upon goods and services. Thus, depending upon their marginal propensity to consume, a proportion of this additional disposable income will be spent on additional consumption again giving rise to the positive multiplier effect. Similarly, if firms are not so heavily taxed, their profits will be higher. Such enhanced profitability may encourage them to undertake projects that were previously deemed to be marginal. For example, the high costs of developing a particular site may have put off a firm wishing to build upon it, however lower taxes could partially offset such high costs, enabling an acceptable profit to be made by going ahead with the development. Therefore, more commercial activity is likely to be undertaken, again stimulating the positive multiplier process.

Therefore, effective expansionary fiscal policy, whether achieved by **increasing government expenditure** or **decreasing taxes**, should enable the government to stimulate the economy out of recession. This should lead to a shift in aggregate demand from AD_0 to a higher level such as AD_2 in Fig. 9.1. Here we would have a fuller utilization of the nation's resources (given current technological and economic conditions) as output would reach Q_2 at an acceptable rate of inflation as the general level of prices has only risen to P_2.

As an economy emerges from a recession, general aggregate demand increases causing the aggregate demand curve to shift to the right. This effect, stimulated by the positive impact of the multiplier process, is likely to lead to an increase in building orders in all sectors of the property market ranging from extensions to new build. For example, higher consumer demand could lead to more people working in the retail sector, offices, and factories, and therefore there is likely to be a demand for more building space to be created as existing space becomes fully utilized. In the housing market we would also expect an increase in activity as the average consumer becomes better off. However, because there is a time lag involved in the production process, due to the time taken from site acquisition to completion, such increases in demand are likely to put pressure on existing space, pushing up rentals and prices as people compete for the available buildings, until new buildings become available.

(b) 'Overheating'

If the economy begins to 'overheat' it is normally because the growth in aggregate demand has outstripped the economy's ability to supply the increase in the level of economic activity. The economy is liable to overheat when it reaches the peak of a cycle (Fig. 11.1), where aggregate demand has reached very high levels such as that shown by the aggregate demand curve AD_3 in Fig. 9.1. Once we have reached the stage whereby suppliers cannot currently cope with further increases in orders, any additional increases in demand are likely to lead to inflation as prices rise, or a balance of trade deficit is produced as imports are used to accommodate the excess demand. In fact, referring to the first point, this situation of excess demand is often referred to as an **inflationary gap**. Inflation is likely to occur as price rises are used as a way of rationing out existing supply to the highest bidder, and an increase in imports could occur as imports could become cheaper and more accessible. Neither situation of high inflation or an imbalance on the trade account is desirable. Moreover, if demand were allowed to increase yet further these problems would increase. Therefore, government could attempt to 'dampen down' such high levels of demand by introducing a **contractionary fiscal policy**. Again the same instruments of fiscal policy will be used, but in the reverse direction.

(i) Decreasing government expenditure

By decreasing government expenditure there is usually a decline in government capital expenditure rather than a reduction in current expenditure. The reason for this preference of cutting capital expenditure rather than current expenditure is that the latter will tend to lead to more job losses than the former and is therefore more politically sensitive. Moreover, many may be unaware of future government spending plans and are therefore oblivious to the fact that they have been shelved or scaled down. However, as the government cuts back on projects – for example an extensive hospital building programme may be reduced to the provision of new hospitals only in the most needy areas – less people would be employed by the public sector, less firms will receive public contracts, and therefore general incomes and profitability are likely to fall. If this does occur, those people who have lost their jobs will spend less in the shops, and thus the retail sector will order less from suppliers. Thus, shops, offices and factories may make people redundant thus exacerbating the problem and creating a negative multiplier. Cuts in current expenditure, say by scaling down the size of the civil service, would also lead to the negative multiplier effect although the impact on the construction industry would not be as direct as the lost orders caused by reductions in capital spending.

(ii) Increasing tax

In order to promote a similar **deflationary effect** to that of decreasing government expenditure, the government could attempt to reduce aggregate demand by increasing tax levels. This policy of raising taxes would leave consumers and firms with lower disposable incomes, and as such they would demand less. Thus, a downward multiplier would again be initiated, as well as a contraction in the magnitude of any positive multiplier effects in the economy.

Therefore, **effective contractionary policy**, whether achieved by **lowering government expenditure**, or **raising taxation**, should be able to pull the economy down off inflationary peaks back to a more steady level of growth. However, it must be appreciated that both contractionary and expansionary fiscal policy can fail. Even if this failure is not total, the policy can either fail to reach its desired objectives in full, or it can produce other problems. One can perhaps draw an analogy here with drugs that are used to cure a disease in a person. Firstly the drug may not be fully effective on its own, and it may therefore need to be used in conjunction with some other form of treatment. Secondly it may have unpleasant side effects that also need to be counteracted. To appreciate this point yet further the book now briefly examines the potential sources of failure that are inherent in fiscal policy.

The limitations of active fiscal policy

The potential limitations of fiscal policy are numerous. However, as we discuss them below, you should bear in mind that methods could be devised to limit, or even eliminate, some of these typical problems.

(a) Time lags

Time lags exist because there is a delay between the implementation of a policy and its final results. In other words, the net impact of a policy may occur some time after the policy and its associated measures have been introduced. Such time lags can be divided into two categories, namely **decision lags**, and **execution lags**. With respect to decision lags, policy may be delayed simply because it takes time to recognize that the economy is in an undesirable situation. For example, it can take six months, or more, to collect reliable statistics. Moreover, before taking action, and deciding upon its magnitude, government must decide whether the upturn, or downturn, of aggregate demand is the beginning of a major change in the economy, or will be just a minor 'blip' before it settles down to its normal growth path again. Execution lags, on the other hand, occur because it can take some time for the full effects of fiscal policy, once initiated, to be worked through. For example, long term spending plans, such as those on the construction of new hospitals, cannot be changed overnight. Moreover, it can take a long time to fully implement such a policy: the hospitals need to be planned, suitable sites found, buildings need to be built, staff need to be recruited, and so on. In fact, this process may only be completed several years after the original inception of the policy itself. Furthermore, an even larger impact on the economy is likely to occur through the multiplier process that such an investment would 'trigger off'. Thus, even further effects may be felt both on a wider scale and at a later stage. Execution lags can be further lengthened as delays created by unexpected shortages, strike action, political wranglings etc., extend the time before the policy can be fully implemented.

Therefore, because of the existence of such time lags two major problems can occur:

1 The defect in the economy that the policy is aiming to cure can be made worse. For example, if government raised public sector expenditure in an attempt to boost aggregate demand and ward off a recession, it may find that other variables of aggregate demand have independently and unexpectedly increased also. Thus, the economy could experience increases in public spending as well as increases in private consumption and investment expenditure. Therefore, the end result is that, because the activities of the private sector were not adequately forecast, the economy has ended up with more than the boost intended by policy makers. In this situation the economy could move from the problems of a recession to the problems of 'overheating'. This could be represented in Fig. 9.1: a government could attempt to increase aggregate demand from AD_1 to AD_2. However, because of the increased and independent activity of the private sector aggregate demand could end up as high as AD_3.

2 If the government felt that the economy only needed a small boost in aggregate demand, in order to prevent it falling below the desired level of growth, it may only sanction a small increase in government expenditure. However, if this rise in public spending was unexpectedly accompanied by a dramatic collapse in investment expenditure from the private sector, the overall impact on the economy could be nullified, or indeed reversed. Therefore, far from aggregate demand moving from AD_1 to AD_2, it could actually drop to a lower level such as $AD_0{}^*$.

(b) Uncertainty

The government faces two major sources of uncertainty when deciding the time to implement fiscal policy, and to what degree. Firstly, it cannot know for certain the value of key variables in the economy such as the marginal propensity to consume. Without accurate data on the marginal propensity to consume, the exact magnitude of the multiplier, for example, cannot be known. Such information can only be gathered from past data and informed forecasts. Mistakes in assessing these values could lead to the aggregate demand schedule shifting by too much or too little. For example, if the government believed that the magnitude of the multiplier was in the order of 2.5, it would expect overall incomes to increase by two and a half times from any injection in government expenditure. Therefore, if, in reality, the multiplier was then found only to be of the order of 1.5 for example, the initial increase in government expenditure would not have been sufficient to reach the desired objectives of the public policy maker. The second area of uncertainty is that, as suggested above, the government is unsure about how much other variables such as private consumption and investment, will have moved independently by the time the policy takes effect. If other components of aggregate demand change considerably the net effect of the policy may be quite different to what was originally anticipated.

(c) Fiscal drag

It has been found in the past that much of the positive momentum created by an expansionary fiscal policy can be reduced via the process of 'fiscal drag'. The problem occurs when consumers' incomes increase, aided by the multiplier process, they are pushed into higher tax brackets. Therefore, the full impact of an increase in incomes is reduced as much of the additional income is then taken back in the form of higher taxation, and thus does not lead to a great increase in disposable income and aggregate demand.

(d) 'Opposing multipliers'

Government must be careful to ensure that if they wish to create growth in the economy, for example, they are not, at the same time, initiating policy which could dampen down such growth. For instance by looking at the 'Balanced Budget Theorem' one can see that the full effect of an injection of government expenditure, and its associated positive multiplier, could be severely reduced by a corresponding leakage and its associated negative multiplier. Such a leakage would occur for instance if taxation were to be raised to pay for the above increase in government expenditure. For example, if the government wished to spend £200 million. on a new road building project, and assuming a marginal propensity to consume of 0.75, the positive multiplier formula would suggest that overall incomes in the economy would be boosted by £800 million. This can be shown in numerical form as:

$$Ku = \frac{dG}{(1 - mpc)}$$

$$= \frac{£200\text{m}}{(1 - 0.75)} = \underline{+£800\text{m}}$$

However, if government were to raise taxation by £200 million. in order to pay for this increased level of public spending, one would also have to calculate the impact of the corresponding negative multiplier caused by the tax increase:

$$Kd = \frac{-mpc\ d\,T}{(1 - mpc)}$$

$$= \frac{-0.75 \times £200\text{m}}{(1 - 0.75)} = \underline{-£600\text{m}}$$

Therefore, the net effect of the policy will be the difference between the positive effect of the upwards multiplier, and the negative effect of the downwards multiplier. Thus, in this case, overall incomes will only have increased by £200 million:

$$\frac{dG}{(1-mpc)} + \frac{-mpc.\,dT}{(1-mpc)} = \frac{£200\text{m}}{(0.25)} + \frac{-0.75 \times £200\text{m}}{(0.25)}$$

$$= £800\text{m} - £600\text{m} = \underline{+\,£200\text{m}}$$

(e) The failure of 'pump priming', and the theory of 'crowding out'

Many advocates of fiscal policy believe that a small increase in government expenditure will lead to an increase in investment expenditure from the private sector. The theory behind this view is that government spending will promote growth in the economy, via the multiplier process, which should stimulate the confidence of investors to expect a more buoyant market. Therefore, as the public sector spends more, businesses will increase their levels of investment so as to prepare for the expected increased sales volume. This notion of stimulating private investment via an initial injection of government expenditure is known as **pump priming**. However, empirical evidence has shown that such pump priming has frequently been associated with decreases in private sector investment expenditure rather than increases. In fact, it is often argued that such a policy 'crowds out' private investment. In its simplest form, **crowding out** can be explained by the fact that if governments undertake major development projects themselves, with their own organizations, there is less of such work available for the private sector to get involved in. For example, a private sector house building firm may shelve plans to build houses in an area if the government decides to build public sector housing in that area. In this instance much of the market will have been accounted for by the new public sector housing, thus making it harder to sell the private houses, and therefore making such a development unviable and unprofitable. In other words the necessity for private sector involvement in the market has been 'crowded out' by public sector activity. However, the more usual form of crowding out is as follows:

If the government borrows the money it requires for increasing its expenditure, this could lead to upwards pressure on interest rates making it too expensive for firms in the private sector to acquire borrowed funds for investment finance.

Interest rates could rise for two reasons. Firstly, if the government borrowed money from the financial institutions, they would be in competition with private borrowers for the available funds from such institutions. As such, the financial institutions could ration out their limited supply of loanable funds by using the price mechanism. This they would do by raising the price of finance (interest rates) so that the loans went to the highest bidder resulting in the financial institutions reaping the highest possible reward (profit). A second possible explanation of rising interest rates would be if the government were to borrow directly from the public. The government could borrow from the public via the sale to them of government bonds for example. In order to persuade the public to put their money in such assets they would need to be offered an attractive return in terms of both security and profitability. If better terms were offered by the public sector, investors may begin to withdraw their money from financial institutions, such as the banks and building societies. In order to maintain business, these institutions would have to attract funds back into them by offering greater incentives than those being offered by the public sector. It is in this way that one could get interest rates increasing due to an initial increase in public spending.

International research has shown that the problem of crowding out certainly exists but it is not total. However, these empirical studies are interesting as many are classic examples of 'poor science'. The reasoning behind this accusation is that many researchers 'fall into the trap', or are obliged to do so, of proving their view rather than merely testing to see if it is correct or not. Thus, there exists an enormous range of statistical 'findings': research from people who advocate a prominent role for fiscal policy in the management of the economy, suggest that crowding out may exist but it is of an insignificant magnitude to be problematic. On the other hand, studies by those who believe in a less interventionist approach to demand management show that the magnitude of the crowding out problem is so great that it could totally nullify the usefulness of fiscal policy.

Conclusions on fiscal policy

As we have seen, there are a variety of potential problems with fiscal policy. In fact, many argue that the adherence to this policy as the 'mainstay' of demand management in the past has actually made the cyclical nature of the economy worse rather than better. However, it is most unlikely that fiscal policy should be completely abandoned as many of the criticisms cited above could be reduced. For example, time lags could be shortened via improved forecasting, having 'off the shelf plans' for public buildings, designating future public sector development land, and so on. All such improvements could lead to the more rapid implementation of public policy. Moreover, as we cannot 'rewind' and 'replay' history, we cannot possibly know with any great certainty what would have happened to world economies if fiscal policy had not been used. For example, if economies had been completely left to *laissez faire* the cycles could have been even greater or more frequent. Furthermore, if one felt that the shortcomings of fiscal policy were too great to use it as a major part of demand management, its role could be suitably reduced to that of 'fine tuning' the economy. Fine tuning is where small fiscal adjustments are made to keep the economy on course so as to achieve a steady state of growth. Such fine tuning can be a continual process, but such measures are often announced on a country's 'budget day' when the chancellor, or finance minister, makes alterations to the tax regime and government expenditure in the hope of encouraging the economy to reach government policy targets. Finally, if one is not happy with fiscal policy, an alternative area of demand management could be tried in isolation to fiscal policy, or in conjunction with it. The alternative is known as monetary policy.

However, before leaving the discussion on fiscal policy and going on to examine monetary policy, it is felt necessary to provide a brief explanation of the concept of **automatic stabilizers**. As the name suggests, these are essentially mechanisms which automatically reduce shocks to the economy, and thus the magnitude of cycles, without direct discretionary fiscal policy whereby governments physically intervene in the management of aggregate demand (as seen directly above). For example, if an economy is experiencing a recession, and is thus in a deflationary gap, incomes and output will decline. However, to counteract this decline, injections of money into the system in the form of unemployment benefits, for example, could occur. Moreover, in an economy with a progressive tax structure, leakages from the economy could be reduced as people pay less tax as they move to lower tax brackets as their incomes fall. Both of these effects could help prevent the economy falling even further into recession. Conversely, if the economy was overheating, and we had an inflationary gap, unemployment benefit payments would fall, and people would move into higher tax brackets. Both of these effects would automatically reduce the potential inflationary effects of such a situation.

(2) Monetary policy

Just as with fiscal policy, monetary policy *is* part of demand management. However, monetary policy is perceived as being less interventionist than fiscal policy as it is conducted through the intermediary of the financial institutions. It is for this reason that it is often the preferred tool of governments that are to the 'right of centre' in political terms. Moreover, in comparison to fiscal policy, monetary policy is seen as being quicker to react and to implement. Monetary policy is a technique that enables government to attempt to manage the level of aggregate demand by using the instruments of: changing the **rate of interest**, and/or, changing the **availability of credit**. Therefore, in summary, monetary policy can be seen to have two instruments:

$$\textit{Monetary policy} = dr \textit{ and/or } dCR$$

where d = change in;
r = the interest rate (the cost of credit);
CR = the availability of credit;

However, it must be noted that monetary policy need not be used in complete isolation to fiscal policy as they can be used in conjunction with one another in order to achieve desired policy goals. In terms of 'the' interest rate, it must be appreciated that this term is either referring to the official government guide rate, or is an average of interest rates in the economy. This point must be made clear as there can be a very wide range of interest rates at any one time for both borrowers and depositors. Now, just as with the debate on fiscal policy, we will firstly examine how monetary policy can be used to influence the level of aggregate demand at different stages of the economic cycle.

(a) Recession

If the economy was experiencing a recession, **expansionary monetary policy** could be introduced in an attempt to increase the level of aggregate demand, and thus the level of incomes and output. Such a policy would hope to achieve this via a rise in borrowing and spending by both consumers and firms. This could be tried by lowering the interest rate and/or, increasing the availability of credit. (For a brief description of the conditions created by a recession please refer back to this title under the topic of 'fiscal policy', see p. 166.)

(i) Reducing interest rates

As many consumers obtain credit in order to finance a proportion of their expenditure, it is likely that a decrease in the cost of such loans would encourage more people to apply for them. As more people obtain loan finance their spending power is increased, and via the multiplier process this is likely to have a positive impact upon the construction industry. That is, more consumption expenditure will tend to lead to a retail boom whereby new retail outlets and shopping areas are required. Such consumer booms can lead to the redevelopment of inner city retailing areas, and/or the setting up of new out of town 'retail parks' on the periphery of the urban area. These increases in commercial activity are bound to filter, in part, to domestic industry, perhaps leading to new factory space being ordered. In this process one again sees the derived nature of construction demand.

There is also likely to be a direct demand for construction in the residential sector as the majority of people finance the purchase of housing through borrowed funds in the form of a mortgage. Thus, if interest rates decline, mortgages would become cheaper, and therefore more people could afford to buy their own home. This would have the effect of pushing up the demand for housing and encourage further developments. (This process would eventually come to a halt as house prices were driven up by the rising demand to the extent that housing became too expensive regardless of cheaper housing finance.) Moreover, if people had to pay less for their mortgage each month, they would have more money left over to afford home improvements and extensions to their existing dwellings. Furthermore, they would have more disposable income for general consumption expenditure. Such increases in general expenditure are likely to filter through to the construction industry via the multiplier process (as described above). Lower interest rates should also encourage firms to carry out more work and investment. The reasoning behind this statement is that many projects that were previously unprofitable, or too marginal to risk, could now be worthwhile due to the availability of cheaper project finance. Even if revenues expected from the projects remained unchanged, their cost would now be substantially lower, assuming that a large part of project finance is borrowed money. Finally, low interest rates could stimulate an increase in the magnitude of the multiplier process by increasing the marginal propensity to consume. The marginal propensity to consume may increase as the marginal propensity to save declines because the return on savings has been reduced via a lower interest rate. This is especially the case if the rate of inflation exceeds the rate of interest. If this situation occurs, the real rate of interest is in fact negative. In other words, the purchasing power of one's money actually declines if it is left on deposit as savings.

(ii) Increasing the availability of credit

Just as with the lowering of the interest rate, the attempt here is to make credit more freely available so as to stimulate aggregate demand. In fact, if any government imposed restrictions are lifted from the financial institutions credit should become more readily available to those consumers or firms that are in a position to obtain it. Furthermore, in the absence of a dramatic change in the demand for loans, as the supply of loanable funds increases, the price of credit (the interest rate) could fall. Thus, such a liberalization of the financial markets is likely to enable more to obtain more loan finance. This should then lead to an increase in the demand for goods and services, thereby creating a multiplier assisted increase in aggregate demand.

Therefore, the economy could emerge from the recession, and experience the benefits described earlier after the section on expansionary fiscal policy. However, as with the case of expansionary fiscal policy, there is the risk of the economy growing too fast so that aggregate demand outstrips aggregate supply leading to problems of 'overheating'.

(b) Overheating

If the economy does begin to show signs of such excess demand, a **contractionary monetary policy** could be used in order to dampen it down. With a contractionary monetary policy government could increase the rate of interest and/or decrease the availability of credit.

(i) Increasing interest rates

Higher interest rates mean that any proposed consumption, or investment, obtainable from borrowed funds, becomes less attractive, as does the prospect of obtaining a mortgage for house purchase. Moreover, those already with a mortgage, or other outstanding debts, will be faced with higher monthly repayments thus leaving them with less income for the purchase of further expenditure. (This is assuming that loans are normally based upon a variable rate of interest, and that a rate has not been fixed for the duration of the loan. This assumption is reasonable inasmuch as although the latter form of loan does exist, the former is still the most common.) Therefore, a negative multiplier effect is likely to be set in motion as less goods and services are purchased by consumers and firms. This, in turn, may force shops, offices and factories, to cut back output and lay off staff. Further damage can occur as high interest rates could reduce the magnitude of the multiplier by encouraging a higher marginal propensity to save, and therefore a lower marginal propensity to consume. Construction firms are adversely affected by such a situation as they are likely to suffer from declining orders. Moreover, their profitability will suffer as they are forced to cut tender prices so as to compete for the little demand that exists. Furthermore, extra costs incurred by unanticipated increases in the interest rate can 'fatally' damage projected cash flows. This situation is often made worse as suppliers press for more rapid repayments as the cost of their borrowing increases also. During such periods, many of the weaker building firms could go out of business, whilst others struggle to survive by taking on marginal projects, or by entering areas such as repair and maintenance, or seeking contracts abroad.

(ii) Decreasing the availability of credit

The government could impose restrictions upon the lending institutions in an effort to reduce their ability to lend money out. If credit became less easy to obtain, consumption expenditure should decline, therefore reducing aggregate demand. However, it should be noted that such restrictions could also impair the supply side of the economy as investment expenditure becomes less attractive. The effects of this policy would be further enforced due to the fact that if credit is successfully limited, interest rates are likely to be pushed up as demand for credit competes for the reduced supply of finance. The construction and property industries would both tend to suffer under a regime of credit restrictions as most activity in these markets is conducted with borrowed funds.

Thus, effective contractionary monetary policy should be able to reduce inflationary pressures and encourage the economy to grow at a more gentle rate, just as expansionary monetary policy should promote the level of growth. However, as we have seen with fiscal policy, there are also some potential limitations of monetary policy. Some of the most important problems will now be discussed.

The limitations of monetary policy

Firstly, one needs to examine whether it is in fact logical, or realistic, to expect consumers to react to current interest rates, or whether a longer term view is actually taken with respect to the cost of finance. Consider the following example of housing. Although high interest rate policy is not designed to impair home ownership, housing is an example of a large purchase that is normally financed to a large degree by credit. If current interest rates were higher than normal this should not necessarily mean that the demand for mortgages, and thus owner occupied housing should fall. The logical rationale behind this statement is the recognition that a mortgage is repaid over a long, normally twenty-five year, period, and that interest rates are likely to fluctuate frequently during the life of the mortgage. Therefore, high interest rates merely mean that the home owner would be temporarily worse off as the level of monthly repayments rose. However, in times of lower interest rates, during the life of the mortgage, the home owner would be better off. Thus, it should be appreciated that periods of high interest and high repayments should be offset by periods of lower interest rates and lower repayments, as interest rates fluctuate around a historic norm. Such interest rate fluctuations can be seen in Fig. 11.2. Empirical studies have shown that this behaviour does occur in reality and that the demand for loans only begins to fall when the interest rate is so high that the initial monthly repayments cannot be made. In fact, when examining the area of personal loans, for purchases other than housing, evidence reveals that even when charged very high rates of interest, consumers still take out loans and vary their consumption expenditure little. Even those who face higher mortgage repayments in periods of high interest rates, can take out a personal loan or a second mortgage in order to maintain or increase their current spending power. If the interest rate is variable on such loans, part of them may well be paid off when the interest rate is below its historic norm, and thus relative 'savings' are made.

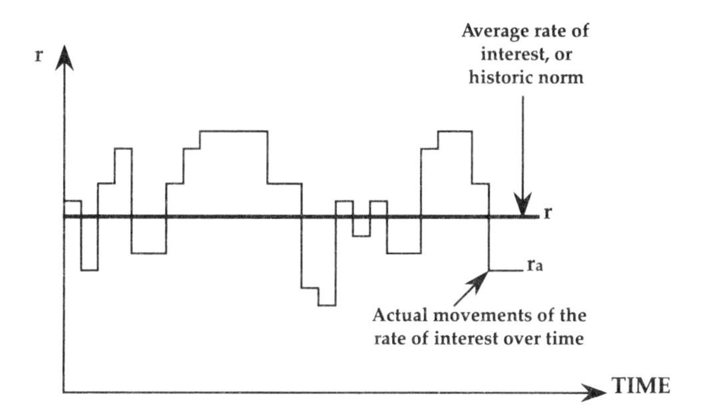

Figure 11.2. *Fluctuations in the rate of interest over time*

A second common criticism of monetary policy is that it is indiscriminate. The argument is that across the board interest rate increases tend to hurt industry as well as consumers. If industry cannot afford to borrow finance for investment, there is a risk that although aggregate demand would fall, and therefore reduce inflationary pressure, aggregate supply would also fall, thus reversing the counter inflationary effect. This problem can be seen in Fig. 11.3: here we have a starting point of an aggregate supply curve given by AS_1, and a high aggregate demand curve of AD_3. High interest rate policy is then introduced which reduces aggregate demand from AD_3 to AD_2, and thus there is a tendency for inflation to fall from P_3 to P_2. However, high interest rates may cause bankruptcies on the supply side, and make new investment too costly. This would have the effect of reducing aggregate supply from AS_1 to AS_2. Once this contraction in supply has occurred the level of choice is reduced in the economy so that even the reduced level of aggregate demand, AD_2, is competing for a smaller number of goods and services, thus driving prices up to P_3. Moreover, such

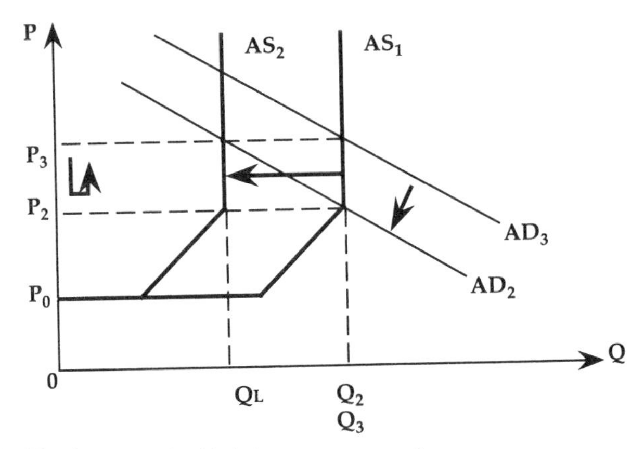

Figure 11.3. *The impact of a high interest rate policy*

a collapse of domestic firms could be an irreversible feature that leads to the country becoming import dependent.

A third problem of monetary policy has occurred with the control of the financial institutions. It has been found that when attempting to decrease the availability of credit, as part of a contractionary monetary policy, many governments have had their aims thwarted by the subsequent behaviour of the financial institutions. Financial institutions who have been instructed to decrease the amount of their lending have found ingenious new ways of redefining or packaging their loans so as to circumvent any restrictions. In fact, some have been known to set up separate lending subsidiaries, perhaps abroad, that are outside the control of the initial policy.

Conclusions on monetary policy

Monetary policy is thus an alternative method of demand management to fiscal policy, although the two are likely to be used in conjunction with one another as an overall package with which to address problems in the economy. However, it can be seen that because of the indirect nature of monetary policy it may be harder to control than the more direct fiscal approach.

A general overview of demand management

The whole principle of demand management primarily rests upon the central assumption that people will react to current economic conditions. Thus, a consumer, for example, will spend according to his or her **current income**, and therefore if government can adjust the level of such income, government should be able to influence the level of current consumer demand. However, this **Keynesian** view of the economy may not be strictly accurate as empirical studies have indicated that people do not necessarily spend according to their current income, but potentially they may make their expenditure plans based upon their *perceived future income*, or even their *previous best income*. If either of these other two points of view hold in reality, the whole impact of demand management policy based upon its Keynesian foundations, will be severely limited.

Milton Friedman put forward his alternative hypothesis known as the **permanent income hypothesis** (*A Theory of the Consumption Function*, Princetown University Press, 1957). Friedman's proposal, which is similar to the **life cycle hypothesis** of A. Ando and F. Modigliani (The life cycle hypothesis of saving: aggregate implications and tests, *American Economic Review*, March 1963), suggests that people's consumption behaviour is related more to their **perceived future income** rather than their present income. For example, Friedman feels that people will 'over' consume now if they expect that their future income will be great enough to offset any debts incurred at present. Therefore, at the beginning of one's working life, one could acquire, via loan finance, a house, a car, household goods, etc., as one could reasonably expect future salary increments and job promotions to pay the debts off in the future. Moreover, ongoing inflation will gradually erode away in 'real terms' a proportion of the debt owed.

Conversely, the **relative income hypothesis** proposed by James Dusenberry (*Income, Saving and the Theory of Consumer Behaviour*, Harvard University Press, 1957) suggests that people's spending behaviour is most strongly influenced by their **best previous income**. Thus, if people's incomes decline, in times of contractionary demand management, or a recession, for example, they will not decrease their consumption by much as they are now accustomed to the standard of living provided by a higher level of income. Therefore, loan finance would be sought, or savings used, in order to defend their previous best consumption position. This could be done in the hope that when incomes rise again, perhaps at the end of the recession or during an expansionary demand management policy, sufficient monies will be forthcoming to pay off the debt. The theory then goes on to suggest that once their original income and consumption pattern has been reached, any further increases in income will lead to an upwards jump in consumption expenditure. This process was termed the **ratchet effect** by Dusenberry. Yet again, the impact of policies designed to influence current consumption may be severely limited by such behaviour.

Due to these apparent inadequacies of demand management some argue that more emphasis should be placed on the supply side of the economy.

(B) Supply side economics

As the name suggests, supply side economics is the study of macroeconomic management that is aimed at 'managing' the level of aggregate supply in the economy. Essentially supply side measures are those that are introduced in order to encourage an expansion in domestic supply. The

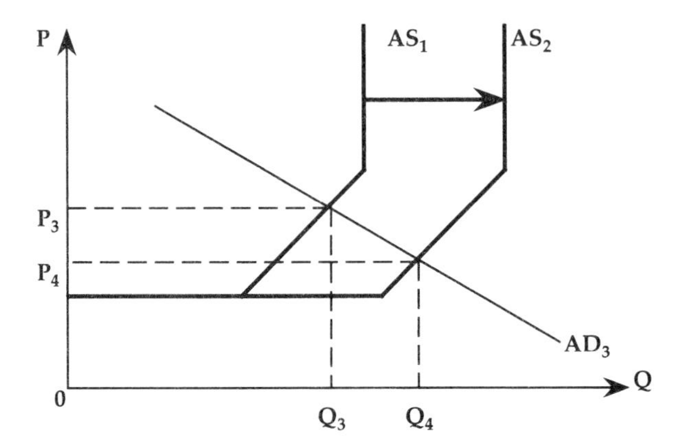

Figure 11.4. *Increasing aggregate supply and reducing inflationary pressures*

underlying logic here is that if you can successfully increase overall domestic aggregate supply, the impact of both a demand led recession, and inflationary periods, in the economy would be reduced. For example, if the economy was growing too fast and 'overheating', as shown by the high level of aggregate demand AD_3, and the aggregate supply schedule AS in Fig. 9.1, prices would be driven up to P_3. However, if domestic aggregate supply could be increased in such an instance, much of the inflationary effect of excess demand could be reduced. This is shown in Fig. 11.4 where increased supply levels shown by AS_2 means that there will be less competition over the available output so that prices only rise to P_4 and not P_3, as output increases from Q_3 to Q_4. Similarly, by promoting circumstances that would enable aggregate supply to be increased throughout the economy should reduce the decline in output, and the level of unemployment, when aggregate demand is low during a recession. For example, if favourable conditions enabled firms to produce more at any given price, the aggregate supply curve could shift to the right as shown in Fig. 11.5. Here it can be

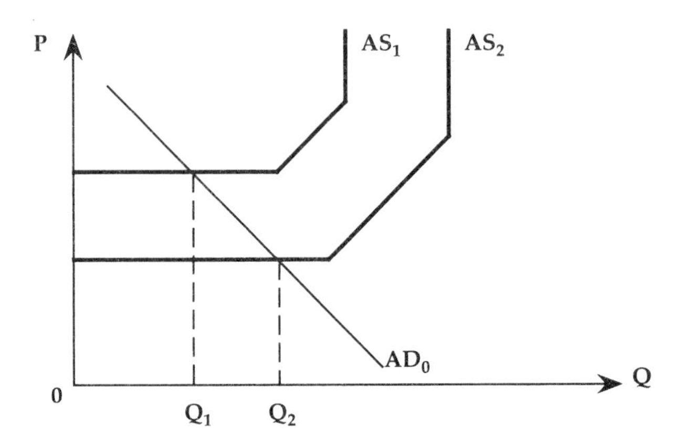

Figure 11.5. *Supply side incentives: a recession*

seen that at a low level of aggregate demand such as AD_0, output would drop as low as Q_1 under ordinary supply conditions shown by the aggregate supply curve AS_1, yet successful supply side policies could shift the aggregate supply curve to the right thus giving a new curve at AS_2, and therefore output may only drop to Q_2. For example, under the old conditions of supply given by AS_1, many firms may have gone out of business thus leading to a substantial decline in domestic output to Q_1. However, given more favourable conditions of supply, via the introduction of positive supply side policies, more firms would have been able to cope with the recession and as such output would not fall as much as previously.

Now that we have introduced the basic principle of supply side economics, we can move on to examine some specific ideas on how government can aim to boost the general level of aggregate supply in the economy. These ideas can be classified under the heading of **supply side policies**, or more meaningfully they should be termed **supply side instruments**, as they are each an instrument of the overall policy of **supply management**.

(a) Income tax cuts

It has been logically hypothesized that very high rates of income tax may act as a disincentive to people to work hard as they will see a large proportion of any additional earnings being taken away in the form of income tax. Thus, lowering the overall level of income tax means higher rewards from employment which should encourage people to get a job, if they do not already have one, or work harder, and earn more, in the job that they already have. That is, more financial reward from work, in the form of higher take home pay, increases the opportunity cost of leisure (as this time could be used to earn yet more money), and should encourage people to work harder. Thus, this policy instrument should increase the overall level of labour productivity.

However, in reality, this process may not occur exactly in the form described in the paragraph above. In fact, one may see the reverse effect occurring: higher financial rewards at work mean that people will have a higher disposable income and could spend this on improving and/or increasing their leisure time. For example, the temptation to join the local golf club, and spend more time there may be enhanced if such an activity now becomes more affordable. Therefore, this potential negative effect on labour productivity could go some way to counteracting the original, positive effect. Moreover, many people cannot actually work more even if they want to, as they are in 'nine to five' jobs and are paid for that period only. In these cases, no reward would be forthcoming to the individual who worked longer hours than this. However, it is probably true to say that because disposable income has increased with such tax cuts, the general morale of the workforce would be improved making people put more effort into their jobs during the official working day. However, this leads on to a further problem associated with tax cuts: if people in existing jobs work harder, as suggested above, more work will be completed in less time than was previously the case. Such increases in productivity could reduce the number of vacancies, or indeed lead to redundancies as the work of some is now done by others, thus leading to increased unemployment. For example a team of bricklayers may normally work on a building site from eight in the morning to six o'clock at night. However, if taxes were reduced they may try to work faster and for longer hours so that they could complete their existing job and move on to another site. If this were to happen the level of competition between bricklaying teams would increase and some would become redundant as others did the work. In fact the benefits of the tax cuts could be quickly eliminated as bricklayers reduced their tender prices in the hope of attracting work in this highly competitive market.

(b) Decreasing the power of trade unions

There are some people who perceive trade unions as being out-of-date organizations that have failed to adapt to socioeconomic change over time and have tried to enforce outmoded restrictive practices within industries. It is claimed that the effect of this is low output, thus leading to a lower aggregate supply than would be the case if industry were allowed to modernize and operate in a completely unrestricted manner. Moreover, many feel that 'closed shop' arrangements, for example, have kept wages artificially high leading to low labour employment, and the non-competitiveness of domestic firms that could lead to their closure in the face of international competition. Therefore, if the power of the trade unions was decreased, or in fact eliminated altogether, the supporters of the argument above believe that more modern and efficient work practices would be adopted by firms, and wage rates would become more competitive. Thus overall domestic supply should be enhanced. However, attempting to introduce such a policy is likely to produce a fierce political debate between those who believe in the good of intervention and regulation, and those who believe in promoting a liberalized labour market. For example, those who advocate the need for trade unions could argue that without trade union support construction workers could lack job security, and be forced to work in a dangerous environment where the emphasis was placed upon productivity rather than the safety of employees.

(c) Decreasing structural and frictional unemployment

Many countries lose a great deal of potential labour output due to both structural and frictional unemployment. **Structural unemployment** is the situation whereby key traditional industries have closed down, or have shed significant amounts of labour, putting a large number of people out of work. This is often a problem as many of such workers are only trained to work in their now defunct, or declining, industry. In order to reduce the levels of structural unemployment it has been suggested that governments should provide re-training programmes so that structurally unemployed labour can start a 'new life' by working, and producing output, in another industry. Similarly, although strictly not under this category, the provision of training for school leavers could decrease the likelihood of youth unemployment by giving them employable skills. On the other hand, **frictional**

unemployment is temporary unemployment caused by people who are moving between jobs. Such frictional unemployment could be reduced by providing labour with assistance concerning such a move. For example, bureaucratic procedures could be simplified and sped up if they are a contributory cause of any delay; or, if the delay is caused by the fact that people need to live elsewhere to take up a new job, their move could be helped along by the giving of a 'relocation package', offering a range of financial incentives. Relocation packages can enhance the mobility of labour by helping with the high costs of moving house – legal fees, estate agents fees, and government taxes on house purchase such as 'stamp duty' (if applicable). To minimize the burden of such policies on government, the private sector could be offered tax incentives so as to encourage them to take on young trainee labour, or set up training programmes. By reducing the number of people who are out of work the nation's overall output, or aggregate supply, should be enhanced.

(d) Reducing unemployment benefits

Many countries operate a scheme whereby if you are in work you pay towards a state insurance programme so that if you or others become unemployed, there will be a pool of money available from which to claim payments until employment is found. Some argue that such a system potentially retards aggregate supply on two grounds:

1 if you can get money for not working, some will not bother to find work;
2 the insurance payments, deducted from your earnings, can be perceived as a tax, thus giving people a disincentive to work.

Therefore, the argument here is to either reduce unemployment payments so that, unless people wish to live at a purely subsistence level, they will need to seek work; or, abolish such a system altogether so that people will have to find work. Unfortunately, however, the obvious faults with such a policy are:

1 there may not be any jobs anyway;
2 there may be jobs but people are not trained to do them;
3 there may be jobs in the prosperous area(s) of the country, but it is too expensive to move there, or there is insufficient, affordable accommodation in such areas.

(e) Profit sharing schemes

Many workers have claimed that they feel 'alienated' at work and therefore owe little allegiance to the firm for which they work. As such, the effort that they put into their work is far from being their full potential. Thus, in an attempt to reduce this sense of 'ill feeling' it has been suggested that governments should encourage firms to introduce 'profit sharing schemes'. The schemes would operate in a way that if output and profits do increase, the employees of the firm would benefit by receiving a 'profits related bonus' to their wages. In the same manner, wages could be cut if output leads to losses being made. Such a scheme could be operated by all forms of firms ranging from surveying practices to building companies. Such schemes have been adopted by some organizations, but with varying success. A common criticism of many profit sharing schemes is that the actual percentage of profits that is distributed to the workforce is so small as to be insignificant, and therefore their total impact upon aggregate supply is likely to be small.

(f) Grants and subsidies
In some instances firms may need an initial incentive to encourage them to set up in business and therefore add to national output. As such, grants or subsidies could be given to new firms in order to help cover the initial cost of labour or capital employment. Indeed, for an existing firm, a grant or subsidy would reduce the current costs of operation so that it may become profitable to expand output and again we would see enhanced aggregate supply. (Refer back to cost and revenue theory, and the theory of the firm in Part 2.) However, in reality, the positive impact of such policies only lasts as long as the policy itself. After the termination of benefits, when grants and subsidies are removed, evidence has shown that we frequently return to the original situation with firms going out of business, or cutting back output, as they cannot survive without public assistance.

A brief overview of supply side policies

As one can see from the above, there are arguments both 'for' and 'against' supply side policies. Which side of the argument one takes in each case could depend upon your political beliefs just as much as your economics inasmuch as many of these points are contentious and emotive issues. Moreover, policies will only be successful if they are introduced intelligently on the foundation of well-researched theory. For example, one must find out whether income tax cuts will lead to people working harder or actually give them an incentive to take more time at leisure. Both these effects are likely to occur and therefore their relative magnitudes need to be assessed so that one can realistically gauge the likely overall impact of the policy. Furthermore, it is unlikely that supply side policies can be used effectively in complete isolation from demand management policies. Therefore supply side economics should be viewed as a complementary policy to demand management rather than a substitute for it. With respect to empirical evidence, many supply side policies have been in operation for such a short period of time that the data are not yet sufficiently significant to prove either their success or failure.

(C) Local government economic policy and the built environment

It should be noted that in the same way that national governments aim to control the national economy, local governments, such as local authorities and town councils can also attempt to control their own local economies. At the local level a variety of techniques can be used to manipulate the level of investment and consumer spending in order for an area to be directed towards a desired economic objective. For example, if a local authority was confronted with the problem of a declining economy, it could aim to stimulate the economy and initiate a corresponding positive multiplier process in a number of ways such as:

1 The giving of low interest loans or capital grants for businesses to set up in the area in the hope of stimulating investment and employment.

2 The granting of subsidies to householders in order to enable them to improve the quality of their dwellings to an acceptable standard.

3 The spending of public monies on a large project such as an urban renewal scheme with the aim of improving local infrastructure and increasing the attractiveness of the location for potential investors looking for a suitable business location.

4 Encouraging firms to locate in the area by offering them a temporary tax holiday from local taxation such as business rates. This would have the effect of lowering the operating costs of the firm in the short run to help counteract high start up costs, and costs incurred in establishing a market.

The above list of policies is by no means an exhaustive one as it is simply designed to give you an insight into how 'macroeconomic policies' can be designed to target the needs of smaller, local economies.

CONCLUSION

It can now be seen that the property and construction industries are inextricably linked with the performance of the (macro)economy in general. Moreover, economic theory enables one to forecast the likely impact upon these industries of government policies aimed at tackling macroeconomic problems at both the national and local level. Thus, the simple analysis presented in this section should enable you to judge, with a fair degree of accuracy, a whole range of future events in the built environment due to changes in the macroeconomy or macroeconomic policy. To suggest just two possibilities: what will happen to the housing market in times of an expansionary monetary policy; or, what will happen to industrial rents when the country is suffering from a balance of trade crisis?

PART FOUR

Urban economic policy and the making of the built environment

The purpose of this part is to examine how economic forces help to shape our urban environments, and how such forces may give rise to the need for some form of public intervention in the urban context. Equally importantly this part of the book will look at a variety of economic techniques that enables one to understand a whole range of urban issues such as the location of buildings and the rationale for town planning. To begin with, Chapter 12 examines the main underlying themes that have produced the towns and cities that we know today. Importantly this part also introduces one to a variety of key terms that can be frequently referred to in any in-depth analysis of the built environment. In recognition of the key role that the private sector plays in the creation and maintenance of urban areas, Chapter 13 goes on to examine a variety of private sector investment appraisal techniques that may be used by the private entrepreneur when making decisions about buildings. Such techniques will be illustrated at the 'micro' level when examining changes to existing buildings such as investing in enhanced insulation, as well as at the 'macro' level when looking at the development of new buildings for example. Chapter 14 then moves on to comparing how such decisions would be taken by the public sector whose overriding goal is perhaps that of maximizing 'social welfare', or the 'well-being' of the urban public at large, rather than that of profit maximization as would be the case in the private sector. To conclude Chapter 15 explores the rationale for government involvement in the urban environment, and how both the public and private sectors can work together to produce towns and cities that are enjoyable to live in, satisfy the needs of their residents, offer plenty of employment opportunities, and are kept as free as possible from the negative aspects of urban living such as pollution, congestion and crime.

— 12 —

The creation of the built environment

Introduction: defining the urban area

Before discussing the main forces that have helped to create the villages, towns and cities that we know today, it is best to try to define what is exactly meant by the term *urban area* so that we introduce a constraint, or boundary, upon our field of study. In 'ancient times' it was normally far easier to define the urban area. This was largely because most people lived a rural life off the land, and larger, concentrated, inhabitations were few and far between. Moreover, because of the need to defend urban areas from a variety of hostile attackers such early towns and cities were often surrounded by defensive walls. Therefore, many urban areas were contained within 'city' walls which gave one a clear demarcation between the urban area and the agricultural land or waste land that lay outside those walls. Although few modern towns that have a historical past will still show direct evidence of such early defences, some do still contain parts of the old city wall although the modern urban area may have spread far beyond its original limits. In the absence of any physical remnants, further evidence of past city walls can perhaps be seen in the nomenclature of present areas of a town or city. For example, some areas are named after a *gate* as this is where a gate used to exist in the historical past. Because of these physically imposed boundaries most early settlements were both **contiguous** and **monocentric**. The term contiguous implies that the built area was a continuous one, and did not consist of any 'satellite' areas or separate suburbs that were split from the main urban area as we may see today. The term monocentric implies that most of the early urban areas just had one centre that was probably a market that served as the catalyst for the original creation of the settlement itself. This situation has changed dramatically in recent times as many urban areas are becoming both **discontinuous** and **multi-centered**. In other words, parts of the town can be separated from other parts by the designation of a 'green belt' for example, and there can be several different centres such as financial centres and agricultural marketing centres to name just two possibilities. Fig. 12.1 shows the main differences between the urban area in ancient times and the urban area in modern times. A further difficulty in defining the classification of an individual town or city is that the expansion of nearby urban areas has frequently led them to eventually join together and form larger **conurbations**. Thus, the classification of a modern urban area is not as straightforward as it may originally seem. Three main classifications can be used:

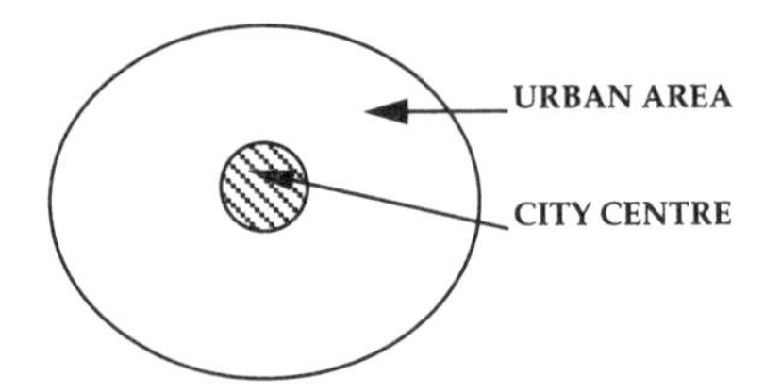

A contiguous and monocentric urban area

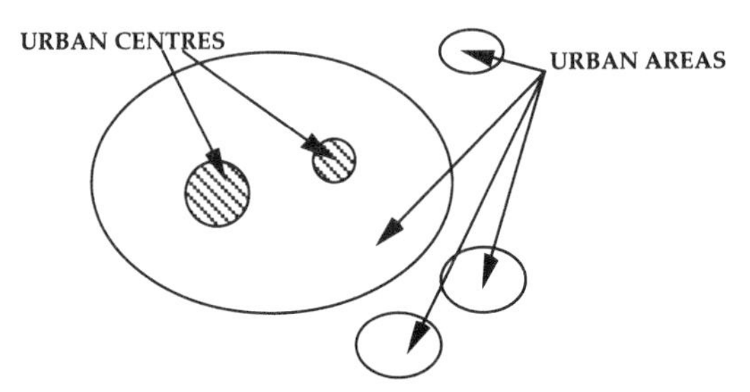

A discontinuous and multi centred urban area

Figure 12.1. *Basic urban structures*

1 Physical definitions: as can be seen from the above analysis such physical definitions may be quite difficult to utilize due to the difficulty in being able to recognize the limits or boundaries of an urban area. However, a 'common-sense approach' may suffice by simply identifying *areas that are dominated by buildings* rather than any other form of land use such as agriculture. The problem with this view is that one would have to use one's discretion and avoid the temptation to interpret guidelines 'to the book' because obviously, for example, a public park in a city is part of that city and should be considered as part of the urban area.

2 Political definitions: for purposes of administration, the provision of services such as sewage and waste disposal, and the raising of local taxes, most urban areas will be officially demarcated into administrative boundaries. Therefore, a study of a particular town can be confined to such constraints. However, difficulties can arise here as towns grow beyond their officially recognized boundaries, or alternatively the administrative boundary may currently include much non-urban land uses as land is set aside in structure plans for future growth.

3 Functional definitions: it has been suggested that urban studies could be involved in any area that 'functions' like a town or city. This definition sounds rather vague at first sight but it enables one to recognize that the urban area may not just be restricted to its obvious concentration of buildings. Rather one can perceive an urban area as being the centre of an *urban economy* that dominates much of the area around it. Agriculture on surrounding land may be largely growing produce to be sold in the nearby town, and therefore the existence of that agriculture can be seen to be largely due to the needs of the urban area.

In order to help us with our analysis there are a variety of **land use models** that predict and explain the likely layout of an urban area, and help to demonstrate how land use patterns in that area may change over time when subject to a variety of economic, political and demographic forces. These models primarily stem from original agricultural land use models as essentially the same techniques and logic are used to describe patterns of agricultural land use and patterns of urban land use. In all of these land use models the concept of the **bid rent curve** is used as the main determinant of actual land use. A bid rent curve represents how much a particular land use, such as retailing for example, will be willing to pay in order to locate in a particular area. As these curves describe the relationship between the price of land and the quantity of it demanded they are effectively the demand curves for the various different forms of land use in the urban area. An important determinant of such demand is the *general accessibility* of the site. Examples of these demand, or bid rent curves can be seen in Fig. 12.2. In order to simplify the analysis for the purposes of making the diagram as simple as possible I have only assumed five categories of urban land use, namely: prestige offices, retailing, middle to low income residential areas, industrial use and higher income residential areas. Assuming that there is a need for 'prestige offices' to occupy a central location in a town or city, such as agglomeration economies, or public image, such establishments are likely to be willing to pay a high price to obtain such accommodation, yet they would be less interested in securing accommodation further away from this area. It is for this reason that their demand curve would fall off rapidly as one moved away from the city centre as shown by the demand curve for such properties D_1. As this demand curve reflects that these businesses are willing to pay more

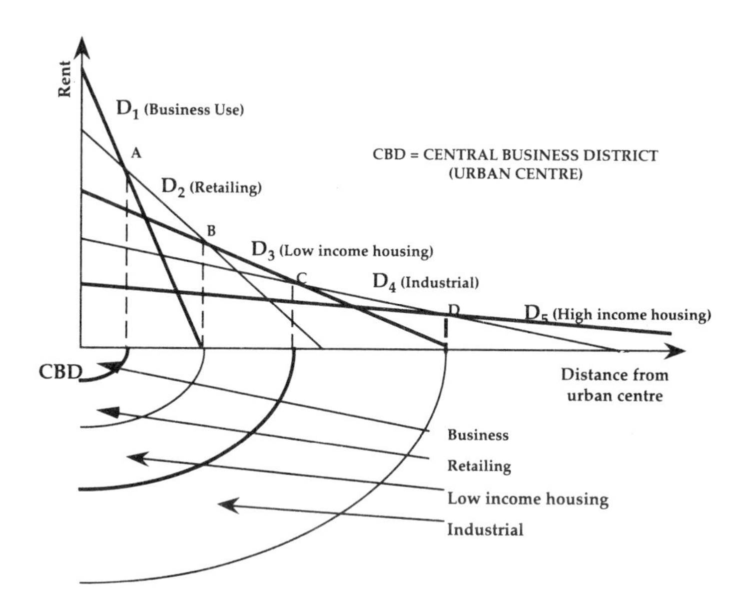

Figure 12.2. *Bid rent curves and the determination of general urban land use*

than any other form of land use to occupy this area, a 'central business district' (CBD) will be created. The boundary of such office use will end, however, when an alternative use is willing to pay more for a location than businesses are. Note that this occurs at point A as it is at this point where retailers, who also desire a relatively central location in order to attract the working populace, begin to out-bid business users as the demand curve for retailing, D_2, is now higher than that for business use, D_1. Just as retailers are displaced when the demand curve D_3 becomes higher than D_2. One can continue this logic to determine the location of all the other forms of land use as seen in the figure. From this one can produce a generalized picture of urban land use radiating out from the centre as seen in the first part of Fig. 12.3. This is certainly not to say that these boundaries will be as clear as suggested in the figure as of course there may be some overlap between different land uses. Some offices may not be able to afford to locate in the central business district, others may feel that there is no need to do so, and some may not be able to find suitable accommodation in that area. It is for these reasons that one will find some offices, subject to planning constraints, in other parts of the urban area. Likewise one may also find residential land use such as luxury pent house flats on the top of high rise buildings in the central business district. Moreover, these implied strict boundaries of land use are bound to be affected by a variety of other issues, some of which are set out below:

1 Geographical features may distort the actual pattern predicted by the model, the most obvious examples being an area of water such as a lake or the sea, or an area of high ground that prevents building. Such geographical constraints are exhibited in the second part of Fig. 12.3.

2 Man-made physical features such as the provision of major transport routes such as roads and railways also tend to distort land use patterns by elongating their location as is also seen in the second part of Fig. 12.3. The reasoning behind this is that land uses become more accessible in

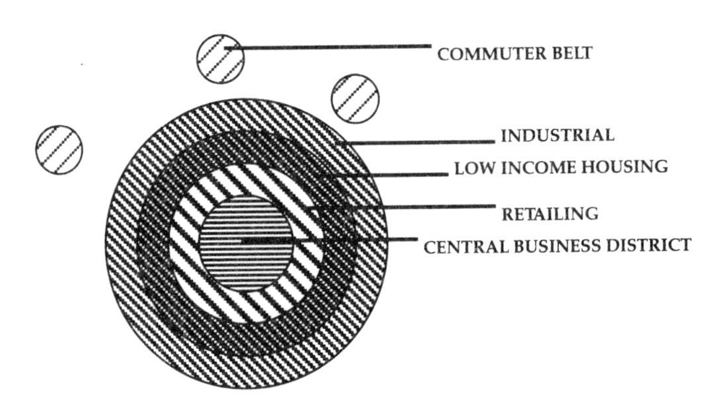

Figure 12.3a. *Concentric zone land use moel*

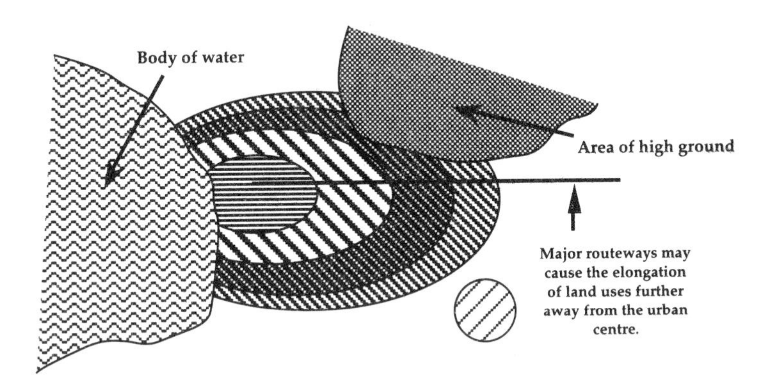

Figure 12.3b. *Concentric zone land use model, the influence of distortions*

terms of travel time and the cost of travel if they are served with good communications.

3 Legislation, such as *land use planning controls*, are also an important force in determining the shape of the built environment and the location of buildings (see Chapter 15). For example, the creation of a green belt is one of the reasons that we may observe high income housing suburbs on the periphery of urban areas as shown in the second part of Fig. 12.3.

4 Historical reasons may dictate that *wedges* of land use appear in towns that break up the general concentric pattern predicted above. For example agglomeration economies, nearness to raw materials, or accessibility to a pool of suitable labour, may have created a noticeable industrial area of a town. Similarly, cultural reasons may produce an area of notably high income housing that does not conform to the normal views of its location. This provides us with a 'sectoral view' of land use in the urban area. Such a sectoral view is shown in the third part of Fig. 12.3.

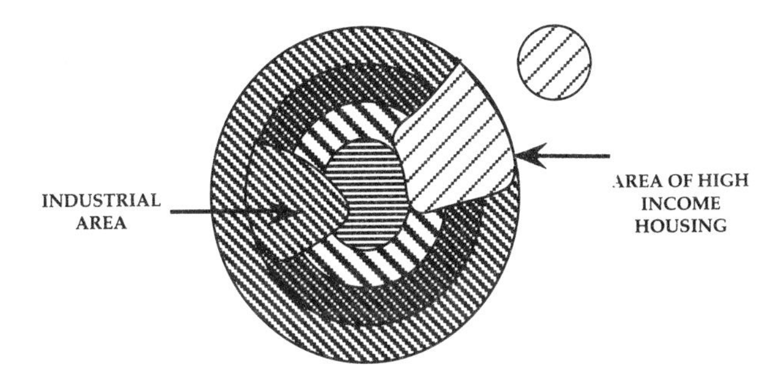

Figure 12.3c. *Concentric land uses distorted by dominant sectors*

Furthermore, one must appreciate that these boundaries tend to change over time due to factors such as the growth of the urban area and changes in transport technology. These forces tend to create a number of *zones of transition* that are undergoing change from one form of land use to another. To illustrate this point it is quite common to see in some cities areas dominated by old, large, houses that were previously the preserve of high income groups, but are now occupied, via multiple tenancies, and the division of such houses into separate units of accommodation, by low income groups, as the higher income individuals have moved out to the suburbs of the urban area as it has grown. Similarly, industrial buildings that used to be on the periphery of the town and close to the old routes of transport, may find themselves 'trapped' well within the city as it expands. Because of problems of accessibility and the high cost of central urban land such industries find it difficult to survive and many towns illustrate the existence of such properties via either derelict central industrial buildings, or ones that have been converted to multiple occupancy. In fact, some imaginative local authorities and private developers have converted such buildings into retail centres or flats. This policy of change in use, refurbishment, and conservation, helps us to retain examples of important architecture from the past, but it also enables one to accommodate changing land use patterns with the minimum of redevelopment and its associated disruption to the local community.

Now that we have examined the basic layout of the urban area this chapter now goes on to look at some of the key forces that are at work in the formation of our urban environment. Specifically the processes of **urbanization** and **decentralization** will be analysed with reference to all forms of property. However, it must be emphasized right at the outset of this discussion that although the processes mentioned here occur in most urban areas there are bound to be exceptions. Moreover, the *timing* of these events is substantially diverse in different parts of the world. For example, many of the modern cities in the third world have grown and undergone substantial transition in only fifty years whereas urban areas in the more industrialized countries often have a longer historical past, and as such these forces have taken place over a longer time horizon. Despite this it is true to say that many countries have urban areas that date back far further than those in the western world, notably there are examples of these in the Middle East. Therefore, it is best to treat each urban area on a case by case basis. However, it is highly likely that the forces now to be discussed will occur at some stage in the creation and development of all towns and cities and therefore they are still important factors to contend with.

(A) The process of urbanization in the formation of the urban area

Urbanization is essentially the process that leads to the formation and growth of urban areas. One can identify a great variety of causes of such urbanization ranging from **economic** to **non-economic** explanations. In the case of the latter category one can identify that many settlements began for reasons of defence, or for religious reasons, or as administrative centres for the government and management of local populations. However, as our main concern in this book is with the subject area of economics it is more appropriate for us to examine the economic reasons for the creation of urban areas in more detail. These economic forces can essentially be split into two major categories, namely **pull factors** and **push factors**.

(1) Pull factors of urbanization

As the name suggests, these are forces that tend to *pull* people into, or attract them to, urban areas. A variety of possible pull factors are briefly discussed below:

1 As subsistence farming is replaced by specialization and there is a recognition of 'comparative advantage' (that is, some farms are in a better situation to produce some forms of output than others), there becomes a need for a central marketing place where produce can be traded. Many towns today owe their existence to such agricultural origins.

2 As industrialization begins to take place, industries tend to locate near one another so as to take advantage of either being close to raw materials, or to benefit from 'agglomeration economies'. Agglomeration economies are essentially the advantages derived from the grouping of plants such as transfers of knowledge and techniques, and the easy use of complementary products, (in other words, one firm's output is another firm's input). In fact, the speeding up of the growth of urban areas tends to accompany the industrialization of a country's economy as people move off the land and go to work in the factories and other related employment of the towns and cities. This migration tends to create a more rapid growth of urban populations in comparison with those in rural communities.

3 Obviously slum areas did exist in the past and can exist today, but as more employment was available in the urban areas it encouraged yet further migration as people wished to take advantage of better living standards, and better security, as well as improved working opportunities. Moreover, urbanization is often synonymous with improved health and hygiene that leads to a decline in the death rate. Therefore, even if the birth rate were to remain constant, or the same as that of rural areas, urban populations will expand at a faster rate than non-urban areas that have a higher death rate. Moreover, research has revealed that the improved living standards of urban areas has often led to a substantial rise in the birth rate, and this coupled with declining death rates has led to a spiralling of urban populations. Such natural population increases are cumulative in nature and are therefore an important cause of the growth of urban areas alongside urban population increases caused by migration.

(2) Push factors of urbanization

Push factors of urbanization are basically those forces that tended to force people, or at least encourage them, to move away from their rural existence. Changes in land ownership in favour of large land owners can force tenant farmers off the land, just as improvements in agricultural techniques, and the mechanization of them, can lead to a lesser requirement for rural labour.

Once urban areas are created by the forces of urbanization their continued growth frequently leads to the process of decentralization.

(B) The process of decentralization in the expansion of the urban area

Decentralization is the process by which there is a movement of population and employment away from the centre of the urban area towards its suburbs. Although it must be said that there can be examples of a decentralization of population, coupled with a continued *centralization* of employment. Therefore, the overall effect is not always that clear. In order to understand the variety of factors that lead to such forces the book now divides the property market into its various parts so that one can fully appreciate the implications for each type of property. To begin with manufacturing property is examined, before going on to look at the cases of residential property, retail property and office property.

(1) The decentralization of manufacturing in the urban area

The movement of manufacturing industry away from the centre of urban areas and towards their periphery can be examined by looking at both supply and demand factors of such industrial decentralization.

Supply causes of manufacturing decentralization tend to occur as there are a large number of supply constraints facing the firm wishing to locate in the inner urban area. Some of these are:

1 There is a limited supply of land in the central area. As such there is a physical limitation that is unacceptable to firms who wish to expand in order to try to meet growing markets.

2 The supply price of land is often very high in the inner urban area because of the demands from other competing land uses that wish to have a central location such as offices and retailing (see Fig. 12.2).

3 The actual supply of existing premises is determined by past economic activity and many of the buildings suffer from functional obsolescence. To convert such buildings, or to completely redevelop the site, would be an extremely costly business, and often the cheapest solution is to find land elsewhere at a more reasonable cost.

4 The supply of land available to industry in the inner urban area has been further reduced by the continued redevelopment of sites for non-industrial use such as can be seen by the growth of leisure facilities in previously industrial areas of some towns.

5 The supply of inner urban land for industrial use is often reduced by the planning system as planners become more aware of the negative impact that some forms of industry may have on the welfare of the urban community. (See the case of **negative externalities** in Chapter 15.)

6 The modernization and increased use of capital in industry has meant that less people are employed by the industrial sector. Therefore, even if industry were to stay within the inner urban area fewer people would be employed by it and they would thus have to look elsewhere for gainful employment.

There are also a number of issues that one could classify as **demand causes** of the decentralization of manufacturing buildings in the urban environment. That is there are incentives for manufacturing to locate on the outskirts of a built area rather than in its core.

1 Agglomeration economies were one of the initial reasons for urbanization and the location of industry. However, it is now argued that such agglomeration economies are becoming increasingly less important in the modern business world. This view is held partly because of the recognition that the 'information technology revolution' (IT) has led to improved, speedier, and lower cost communications irrespective of location. Furthermore, the modernization of transport infrastructures has also led to similar advantages with respect to transport. Therefore, the need for manufacturing industry to be located together is perhaps not as strong as was originally the case. Moreover, much modern industry is classified as 'light industry' and is largely 'footloose'.

2 Modern manufacturing industry has a demand for peripheral locations because the production processes of today tend to require more land. Modern techniques of production normally take place in large single storey buildings as these structures are best suited for long production lines and conveyor belts, large and heavy machinery, and forklift trucks. The multi-storey factory units of the late nineteenth century are not suitable for any of these aspects of the production process, and it is usually these older industrial buildings that tend to dominate the inner city in older towns. In addition to this, modern factories are likely to need more surrounding space for employee car parking and heavy vehicles access.

3 The construction of ring roads and major trunk routes that pass nearby to urban areas, coupled with the decline of rail freight in many countries, has led to an increasing demand for such outlying sites. These new routes tend to be less congested and therefore make the factory more readily accessible.

4 Some commentators on the relocation of industry have suggested that senior management have expressed a preference for such decentralized sites as it makes them nearer to their homes and therefore easier to travel to. Or, perhaps more realistically, such 'green field' sites are seen as a way of enhancing corporate image.

(2) The decentralization of residential property

A noticeable feature of many modern towns and cities is the growth of residential areas at the periphery, or suburbs, of such urban areas. This is known as the *suburbanization* of residential property and such areas are often given the generalized name of 'suburbia'. It is important to note that housing in these areas tends to be of the middle to high income category, leaving the less well off in society to occupy housing in the inner city. It is for this reason that it is quite commonplace to observe housing in the inner urban area that used to be occupied by one rich family now to be divided up into several units and multiply-occupied by people on lower incomes. In fact many of the 'slum areas' of modern cities were originally fashionable districts. Despite the fact that there is, of course, still residential land use within the central areas of the urban environment, large scale decentralization has occurred for such reasons as:

1 There has been a general expansion of urban areas due to the process of urbanization and growths in population. Often the only place to locate new areas of housing for such a growth in numbers is on the periphery of the urban area as there is no space to accommodate them within the existing built area.

2 As a country becomes more prosperous and incomes rise, people tend to demand larger, more spacious dwellings that afford them more privacy. Moreover, outlying areas tend to be more pleasant environmentally, and are away from the negative aspects of the inner urban area such as congestion and pollution. Because of the high cost of central land, such attributes can only normally be gained in the suburban areas.

3 Income growth also tends to enable people to afford longer commuting distances as they can bear the greater cost of travelling to and from their place of work. This, coupled with technological improvements in transport, improvements in infrastructure and the increase in car ownership, has led to the increasing accessibility of suburban residential areas.

It is argued, however, that such a desire for housing at the periphery of an urban area may be checked by the following forces:

1 As people move to outlying areas of a town or city it naturally creates a higher demand for housing in these areas. In the absence of substantial increases in supply at these locations, such increases in the level of demand will tend to push house prices up. Increases in house prices may be sufficient to dissuade people from moving out of areas closer to the centre, but in addition to this the price of housing in the inner urban area

is likely to drop, and therefore make such central properties increasingly attractive.

2 Coupled to the analysis in (1) above, one can observe that some towns have experienced a small reversal to the trend of residential decentralization in the form of **gentrification**. Gentrification is the process by which middle to high income people move back into inner urban areas and purchase older, low income housing, or occupy old industrial buildings, that have been converted to residential use. Reasons for such a move back into the central area are numerous but some of the main causes are:
(i) Inner city residential properties become very reasonable buys as the main housing demand has shifted out to the suburban areas. This relative drop in demand for central housing has been accentuated by the decentralization of other forms of property such as industry. As manufacturing has moved out to peripheral locations, workers have needed to try to follow such jobs so as to avoid lengthy and costly 'inverse commuting'. That is if they do not move they have to commute from the town centre to its outskirts. Therefore, relative falls in central housing prices means that people can purchase a large inner city house, and still have enough finances left to substantially improve that property to their desired levels of quality.
(ii) As countries reach a high level of development families tend to have less children. Therefore, as the size of the family unit begins to decline there is less of a need for the larger houses and space offered in suburban areas.
(iii) Another aspect of highly developed economies is the alarming increase in the divorce rate. This high incidence of marriage failures leads to an increase in the number of 'single' households whereby only one person occupies a dwelling. In such circumstances large housing units are likely to be inappropriate.
(iv) Modern society has led to the increased participation of women in work. Therefore, if both partners in a marriage need to travel into the inner area to go to their places of work it makes sense for them to live closer to work, and to share transport.
(v) There is an increasing awareness of the process of conservation which highlights the possibilities that an old building can present via conversion or refurbishment. Indeed occupancy of such properties can become highly fashionable thus increasing the level of demand for them.

This process of gentrification has, in some cases, created an influx of highly paid executives into areas that were traditionally dominated by low income groups. Moreover, this process tends to be cumulative because as people on higher incomes purchase such properties and renovate them the area is seen as being 'improved', and such improvements are likely to attract even more demand into the area. As demand increases, house prices become inflated, and such high house prices can push the cost of housing beyond the reach of the lower income groups who used to dominate the area. Therefore, due to the process of gentrification, one can witness a change, or reversal, in the social structure of the urban area.

(3) The decentralization of retailing property

The decentralization of residential population, especially those in the higher income brackets, gives an impetus to the decentralization of retailing as the retailers' market moves location. Such a move in the location of the market is backed up by research that reveals that people make the majority of their purchases near their place of residence, rather than near their place of work. Moreover, the increasing emergence of 'car orientated' suburban shopping centres yet further weakens the position of outlets in the central area. Furthermore, both retail and wholesale outlets are keen to take advantage of the cheaper peripheral land in order to construct the larger consumer orientated shopping areas such as the out of town retail centres which are becoming increasingly common around modern urban areas.

(4) The decentralization of commercial buildings

The impact of the forces of decentralization on the office and services market are less clear. There is perhaps still a need for many functions that require direct contact with the public, such as estate agents, to locate in an area of high population density. Thus, a central location is still of importance. However, it is an observable phenomenon that large 'head offices' have tended to decentralize as again the opportunity of cheaper development land is an attractive proposition.

Some conclusions on the process of decentralization

As can be seen from the points discussed directly above, there are strong market forces at work that encourage the decentralization of a large number of land uses for obvious economic and social reasons. However such shifts in the location of urban activity can also be seen to produce negative 'spin offs'. Detrimental effects of decentralization include the problems of 'urban sprawl', and 'inner city decline'.

(a) The problem of urban sprawl

Urban sprawl is essentially the spread of a town or city via low density continuous development. This spread of an urban area can be accentuated by the *ribbon development* of the buildings along major route ways. (A theoretical example of this can be seen in the second part of Fig. 12.3.) Moreover, the forces of growth and decentralization can be so powerful that they are difficult to contain. Even the creation of a 'green belt' by town and country planners, in order to restrict the growth of the built area within certain limits, may fail to achieve its objective as development 'leapfrogs' the newly designated green area. Such a 'jump' in the location of development can be seen in Fig. 12.4. Note that as compared with the initial distribution of land values before the intervention of the planners, land values now tend to rise on either side of the green area as the proximity of the green belt makes location there more pleasant, and therefore more sought after by a variety of land uses, especially high income residential land use. Problems created by such an expansion of urban areas are numerous, but some of the more important considerations are briefly discussed below:

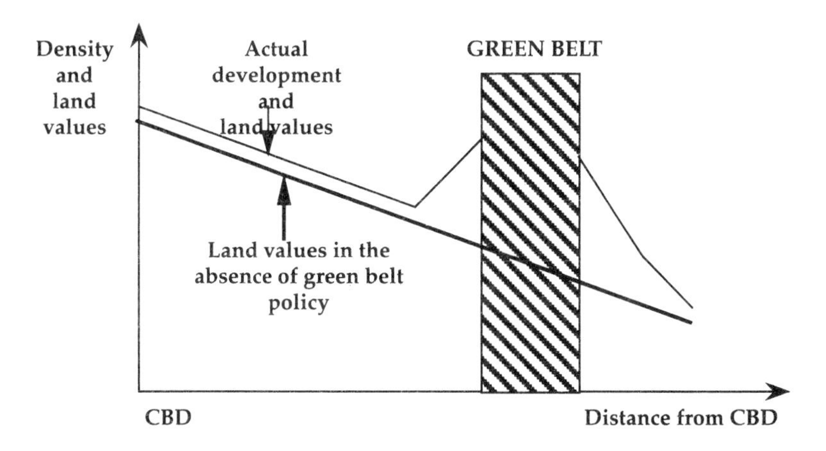

Figure 12.4. *The impact of 'green belts' on urban development and land values*

(i) Some urban land economists argue that urban sprawl is both inefficient and costly, and in particular, these costs are not necessarily borne by the *suburban* dwellers themselves. In other words, it is more inefficient to provide public services, (such as transport), electricity, telephones, gas, water, sewage disposal, and postal services for example, to areas that cover a wide geographical area rather than those urban areas characterized by more high density, concentrated development. If anything, it can be viewed as an extravagant waste of scarce resources. Research conducted in urban areas in the 'developed world' has indicated that the cost of provision of some services has been increased by up to twenty-five per cent by urban sprawl. However, although these costs are passed on to the consumer, the fairness of any 'blanket' increases in service charges is questionable. For example, if the price of electricity for residential users were to be increased because of the greater costs that the electricity company have to meet in order to 'wire up' a larger urban area, these price increases are likely to affect everyone whether they live in the suburbs or the inner city. Therefore, the suburban dweller is effectively being subsidized by the inner city dweller. The inequity of this result is further compounded by the fact that, apart from quirks in the pattern of land use caused by factors such as gentrification, the suburban dweller is normally in a higher income bracket to those who live in the inner city. In other words, low income groups are made even worse off because high income groups wish to locate in the more pleasant outlying areas.

(ii) Urban sprawl can also be viewed as the creator of a variety of general negative effects. For example, the growth of towns and cities means that more green areas are lost and such environmental damage may be irreparable. Green areas not only provide pleasant places for people to relax in, but they are often the important habitats of plants and other animal life. However, one need not automatically assume that all areas that are not currently built upon are positive assets in which there is a need to protect them by environmental or green policies. Some areas can be unsightly wasteland, for example, that can actually be improved by the careful landscaping that accompanies much of the further urbanization we see today. Moreover, much of the land could be currently under agricultural use and therefore not accessible to the general public other than in a visual manner. In fact, an argument against the imposition of green belts in general and the more careful consideration and allocation of 'green wedges' has been proposed due to this view. A *green wedge* would be an area of land that was genuinely an area of outstanding natural beauty, contained the habitat of endangered species, or was accessible for the recreational use of the urban population.

In conclusion, it is felt by many that urban sprawl leads to an increase in both private costs (as in the case of the cost of the provision of services to outlying communities), and an increase in social costs (as exhibited in the loss of pleasure derived from green areas).

(b) The problem of inner city decline

Inner city decline has occurred in many urban areas where the main sources of economic activity have shifted away from the centre due to the forces of decentralization. Such inner city decline is frequently characterized by areas of poverty, coupled with poorly kept and even derelict buildings. As industry vacates the centre of a town to take advantage of a more prosperous economic environment elsewhere, it will leave behind its previous premises. Despite some notable exceptions of successful changes in use and conversions, many old, inner city industrial buildings are not suitable for modern industry, or any alternative use, and therefore, in the absence of the redevelopment of the site, such buildings will remain derelict. Moreover, the likelihood of the redevelopment of the site for an alternative use such as retailing is slight in an area of low economic potential. Such dereliction gives rise to an unsightly urban environment, and such an environment is often blamed by sociologists to create a whole plethora of social problems such as vandalism, and at the extreme, civil disorder as demonstrated by the rioting that has occurred in many modern cities in recent times. In fact, this situation tends to lead to a downwards spiral of a lack of economic prosperity – the area becomes poorer as fewer jobs are available as more firms vacate the depressed location.

Therefore, one can identify a variety of key problems associated with inner city decline:

1 As indicated directly above one of the main problems of inner city decline is with respect to employment opportunities. Many depressed inner cities contain areas that exhibit degrees of unemployment that are substantially higher than elsewhere in the economy. Due to the decline of inner city manufacturing, employment opportunities in the inner urban area have obviously decreased as industry locates, or relocates, in the new suburban locations. Therefore, as inner city incomes decrease, retailing also has more of an incentive to move out to the relatively more prosperous periphery. However, unskilled and semi-skilled workers are often too poor to move out of the city, and therefore they remain, creating an excess supply of labour, or in other words, unemployment. In fact, some cities have shown patterns of 'inside out commuting' as unskilled and semi-skilled workers commute out to work in the industry of the suburbs, whereas the suburban dwellers commute into their offices in the centre.

2 Linked to the above analysis in point (1) directly above, it is found that the inner cities tend to be dominated by residents on low incomes. Because they are on low incomes they cannot afford expensive inner city property prices unless they live in either high density housing units such as terraced houses or flats, or they live in other buildings which are extensively subdivided or multiply-leased. Such intensive forms of living tend to be partially responsible for a variety of 'social ills' such as poor educational achievement and a high incidence of crime. Moreover, conditions are made worse as few can afford essential repair and maintenance of their properties, nor can they meet the cost of improvements that are deemed essential to bring the quality of such housing in line with modern standards. Furthermore, matters are not helped as these residential areas can frequently be found adjacent to old, unsightly, and possibly derelict, commercial buildings.

3 In the original process of urbanization, what are now the central areas of towns and cities were obviously the first parts of the built environment to be developed, and it is for this reason that many of the buildings found in this area are relatively old. Partially because of their age, and partly because of the forces of decentralization, such buildings tend to suffer from functional obsolescence and are of poor economic potential. Therefore, apart from keeping some buildings for purposes of historical conservation, much redevelopment is required to reflect changing land use patterns in the core of the urban area. As extensive redevelopment occurs this can create much inconvenience to those who live and work in central locations due to continual construction and maintenance activity. Construction work can necessitate the diversion of traffic, cause traffic congestion, and create a fair degree of disturbance in the form of noise pollution. Such redevelopment in the heart of the city and the development of new suburbs on the outskirts of the city can lead to 'low points' of construction activity in inner

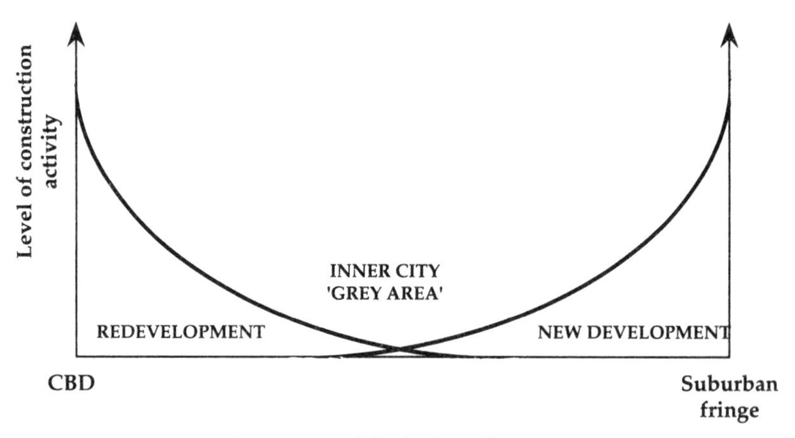

Figure 12.5. *Typical development activity in the urban area*

city *grey areas* as seen in Fig. 12.5. In this figure it can be seen that although people living in the central and outlying parts of an area may be disadvantaged by a lengthy or near continual redevelopment process, they should benefit from newer and improved buildings and urban facilities, whereas those in the 'inner city grey area' will see little replacement leading to the risk of a deterioration in the quality of the building stock in this area.

4 Due to the historical nature and growth of their transport infrastructure inner urban areas often suffer from traffic congestion as routeways cannot adequately cope with the pressures of modern transport demands. The location of roads is largely determined by the functions of the area in the past and the buildings which they serviced. As buildings are normally replaced in a piecemeal fashion rather than *en masse* (except in the case of the *comprehensive redevelopment* of urban areas) such transport infrastructures largely remain intact. Heavily congested roads lead to negative effects such as air and noise pollution, increased risk of road traffic accidents especially involving pedestrians, longer journey times due to traffic congestion, and so on. Public sector intervention can attempt to reduce these problems by the creation of one-way systems, and the creation of pedestrianized zones, but few towns or cities seem to have arrived at a terribly satisfactory solution to the problem, (see Chapter 15, and the debate on negative externalities and road transport in the urban environment).

A conclusion on the creation of the urban environment

As one can see from the above information about the shaping of our urban areas, the urban environment is a dynamic one rather than a static one as it is constantly 'bombarded' by forces of change. Whether these pressures are economic, social, or induced by the public sector, it is important to realize that they can create both negative and positive effects. In the light of this view the remaining chapters examine how both the private sector and public sector operate in isolation and together to form the urban areas that we know and see today.

13

Private sector investment appraisal techniques and the urban environment

Introduction

The private sector is perhaps the leading player in the creation of the built environment in most modern economies. Although it is the public sector which normally controls, or places constraints on, urban decisions (via planning regulations and building controls), it is the initiatives from the private sector in terms of the setting up of industry, offices, or retailing, and even the creation of marketable residential areas, that usually provides the main impetus behind the growth of the urban area. Therefore this chapter examines a variety of common private sector investment appraisal techniques so as to demonstrate how entrepreneurs can make decisions concerning buildings and their location. These decisions can be at either the 'micro level', or at the 'macro level', but the common factor of both is that the private investor is likely to be most interested in maximizing the returns from any investment, or, in other words, maximizing profit.

At the **micro level** the decision maker would wish to utilize a suitable investment appraisal technique that enables them to look into the possibilities of making profitable changes to *existing buildings*. For example at the micro level one may wish to examine whether:

1 A building's operating costs could be reduced via the investment in, and subsequent installation of, *energy saving devices* such as roof or cavity wall insulation or double glazing. If the running costs of a building could be reduced this will increase the gap between revenues received from rent and general costs of the building so that profits will rise. Moreover, the incorporation of such devices may also enable the owner of the building to increase rents as the building becomes a more comfortable and attractive place to work in. For example, with the installation of double glazing not only does the building become warmer and therefore more pleasant in cold weather, but such additional glazing also tends to insulate the building from exterior noises such as traffic congestion. Therefore, in the absence of interference from the public sector, it is sadly highly probable that the built environment will only become more energy conscious if it is profitable for entrepreneurs to become so. Remember that energy savings would also lead to other positive environmental 'spin offs' such as less pollution created by the generation of power.

2 Profits could be encouraged by a *change in use*. The owner of an old industrial building may find that because of the building's age, and its unsuitability for most types of modern manufacturing, it can only attract low paying tenants or multiple occupancy if left in its current use category. Therefore, if planning permission were obtainable, the building could undergo extensive interior re-conversion in order to change its use into an alternative format such as the creation of modern executive flats. In the developed world there are numerous examples of this especially in converted *waterside* industrial buildings such as those found in old dockland areas. This has the advantage of conserving older buildings, or at least their façades, rather than replacing them with new ones, and thus helps to preserve a country's architectural and historical past.

3 Profits could be enhanced via the extensive *refurbishment and modernization* of an existing building. An ageing office block may lack sufficient amenities with which to attract high rent paying occupants. However, if the building were to undergo extensive refurbishment and its facilities were improved the owners could insist on an increase in rents to reflect the higher level of provision, or otherwise seek to attract new, higher paying tenants.

Alternatively investment appraisal techniques can be used at the **macro level**. At this level the investor would be concerned with finding and using the correct investment appraisal technique in order to make larger decisions about buildings than previously examined above at the micro level. For example at the macro level one would be involved in issues such as deciding whether to *redevelop* a site by knocking down the existing building and replacing it with a new one. Obviously such decisions are involved with much greater amounts of money than was the case at the micro level and therefore they need to be taken with even more care and precision.

Whatever the level of investment under consideration, the individual, or individuals, making the final decision to go ahead with the investment or not must address some fundamental questions. Essentially one needs to assess whether:

1 The investment is profitable or not. In other words if the investment is not a profitable one it is not worth undertaking by the private sector.

2 The investment is *the* most profitable option available. Obviously investment funds are not unlimited and therefore if there are several projects to chose from one needs to try to ascertain which one, or ones, will give the greatest financial reward.

Coupled with this basic common-sense 'checklist' of investment desirability (directly above), the investor also needs to ensure that the projects under consideration are affordable, but just as importantly it must be recognized that all projects will carry a degree of **risk**. The existence of risk and uncertainty implies that one will never be 100 per cent sure of the outcome of your decision. For example, the redevelopment of a site may incur unexpectedly high increases in building costs due to unforeseen construction difficulties such as the discovery of uncharted mine works, or methane gas. Similarly anticipated lettings and rentals for the completed building may not be achieved if the economy and property market were to experience a recession. The degree to which risk and uncertainty is considered will depend upon whether the investor is a *risk taker* or is *risk averse*.

Once one has addressed these basic issues concerning the viability and profitability of a project, one needs to decide which investment appraisal technique to use. The choice of investment appraisal technique is absolutely critical as each technique is best suited for a particular investor requirement as will shortly be shown in the following analysis. Essentially investment appraisal techniques fall into two broad categories: **conventional techniques**, and **discounting techniques**. Both of these categories will now be discussed in order to demonstrate their applicability in a variety of circumstances. In order to assist in the explanation of these techniques a simple numerical 'case study' shall be used, the details of which now follow.

Imagine that there are two potential investment opportunities available to the developer: Building *A*, and Building *B*. Also imagine that both of the investments are deemed to be similar in that they both involve the construction of buildings with identical *forecasted* total development costs. These costs could include costs in the acquisition of the land, demolishing any existing structures, clearing the site, and erecting the new building. The similarity of these investments is compounded by the view that, at least at first sight, their returns should be similar. In order to simplify the analysis yet further, it has been assumed that the anticipated operational 'project life' is only four years. This assumption is somewhat unrealistic in the case of buildings which are normally characterized by far longer economic or physical lives. However, the assumption is necessary for the purposes of a simplified explanation and in no way affects the accuracy of the investment appraisal techniques discussed. Furthermore, the hypothetical anticipated income flows generated by the subsequent letting of these buildings on completion exhibit exaggerated differences and fluctuations. Again, however, the figures have been selected to highlight various points and will not hamper the accuracy of the analysis. In fact it would be a useful exercise for you to generate some real figures of your own and insert them into the various techniques suggested to see how they can be used to examine real-life issues. The table of total development costs and anticipated income flows for our two hypothetical buildings appears immediately below where all figures are expressed in £000s.

The figures for net annual income flows would be derived by subtracting the running costs of the building away from the receipts gained from the letting of its floor space. Operating costs include general expenditure upon repair and maintenance, and buildings insurance. Note that the anticipated income stream from building *B* is uniform, yet the income stream for building *A* is anticipated to be high at first and then to drop substantially in the future, although the future income flows are higher for building *A* in the *short run* than they are for building *B*. Such a variation in potential income

Table 13.1 Comparison of costs

Year	Building A	Building B	Description
0	(2500)	(2500)	Initial development costs, (-)
1	1500	1000	Net income flow, year one
2	1250	1000	Net income flow, year two
3	250	1000	Net income flow, year three
4	250	1000	Net income flow, year four
Net	3250	4000	Gross future income flows
Gross	750	1500	Net future income flows

is obviously exaggerated as the time scale of the building's life has been compacted considerably, but a tendency for such fluctuations, although on a less dramatic scale, could be explained by a variety of factors. Future net returns could suddenly 'tail off' as it may be known that the type of building constructed requires significant and costly remedial work at some future date, or it is known that another newer building will be constructed and create competition in the future that could take away the existing building's high paying clients as they move into the more modern building. All the figures in the table would be *estimates* as they involve predictions of future events such as changes in the value of money due to changes in the rate of inflation, or changes in the demand for the buildings themselves.

(A) Conventional techniques of investment appraisal

Conventional techniques of investment appraisal are commonly used, especially at the micro level. Their popularity primarily stems from the fact that they are both simple and straightforward to calculate. However, they are perhaps too simplistic when one is involved in the investment of larger sums of money at the macro level, mainly because they do not strictly take into account the *time value of money* in their calculations, (see discounting techniques later). The two conventional techniques considered by this text are the **Payback Period Method**, and the **Average Rate of Return Method** (ARR):

(1) The 'payback period' method

The payback period of a project is essentially the time that it takes for the investment to recoup its initial cost, and the payback period method of investment appraisal simply recommends projects that pay you back quickly. In other words, using this method, the quicker that a project recovers its initial outlay the better the project. Using the example in the table above, the payback period method would recommend that the developer invested in building *A* rather than building *B* as one would recover one's investment in under two years with the former building, yet it would take two and a half years with the latter. This preference can be expressed as:

$$PB_A > PB_B$$

One glaring potential defect of this technique may be seen in that it really just recommends the liquidity of an investment, and fails to take a longer term view. It has ignored the fact that over the buildings' whole life, the *net* income flow of building *B* is in fact far greater (double) than the net income flow of building *A* (see Table 13.1). At first, this criticism would seem to be so damning as to negate any further debate on the technique, however, there are still a number of advantages of using this method of investment appraisal besides its attractive simplicity. For example:

1 The payback period method of investment appraisal is useful for those investors who need their money back as quickly as possible. If the value of the investment is rapidly recouped two advantages are immediately obvious:
   (i) Investment monies can be re-used to fund yet another development project as there is a relatively healthy cash flow.
   (ii) Any borrowed investment funds for development finance can be repaid quickly. Not only does this save on interest repayments, but it is liable to boost the confidence of the lending institution as they witness liquidity and prompt repayment.

Therefore, this method begins to recognize that monies received now, or in the near future, are worth more to the investor than money received at some distant date. In other words, there is an *opportunity cost* involved in delayed returns.

2 The method is useful when applied to areas that are subject to change such as *rapid technological advances*. The reasoning behind this statement is that if one takes a very long-term view of an investment and its returns, the project may not be able to recoup its investment costs if it suddenly becomes obsolete. Such obsolescence could end the project's prospects of future income flows altogether, or at least substantially reduce them as better, more modern, equipment or facilities are likely to be available to replace any redundant technology. For example, one could design part of a building and have it fitted out so as to accommodate a particular type of machine or computer system only to find that the machine or computer system has become technologically redundant. If this were to occur one would have to look again at one's internal building requirements. Indeed some buildings themselves could become technologically redundant soon after completion if they are of a specific design to meet a particular need.
3 The payback period method is a useful technique in times of *uncertainty*. If future returns are unsure, perhaps due to either economic or political instability, the quicker one can recoup one's investment the better, especially as the longer one attempts to predict net income flows into the future, the more uncertain one's estimates become. This is especially the

case with buildings which tend to have a long economic and physical life. If one were to examine the payback potential of installing double glazing as a means of increasing the insulation of a building, one would calculate the expected payback period based upon current costs of installation, and savings made against current fuel costs. Although *estimates* of future inflation in fuel costs could be made and incorporated into one's calculations, one would not be able to predict future uncertain events such as an *unanticipated* energy shortage, or even the discovery of a new commercially viable energy source. Both of these future outcomes could make any very long term calculations virtually meaningless as the price of energy itself changes.

(2) The 'average rate of return' method

The average rate of return (ARR) merely tells you how much of your investment you will recoup each year, on average, whereby the project with the highest average rate of return will be recommended. The formula used to calculate the ARR is:

$$\text{ARR} = \frac{\text{Gross project returns/initial cost of the investment}}{\text{Life of the project}} \times 100$$

The answer is multiplied by 100 so as to express it in percentage form. Using the figures in our hypothetical case study in the table above one can calculate the average rate of return for both building *A* (ARR_A) and building *B* (ARR_B) as seen below:

$$\text{ARR}_A = \frac{3250/2500}{4} \times 100 = 32.5\%$$

$$\text{ARR}_B = \frac{4000/2500}{4} \times 100 = 40.0\%$$

Therefore, on average, investing in building *A* will give an annual return to the investor equal to 32.5 per cent of the initial investment, whereas building *B* will return 40.0 per cent of the initial investment on an annual basis. As building *B* has a higher ARR than that of building *A* this method would recommend investment into building *B*. This preference for building *B* can be expressed as:

$$\text{ARR}_B > \text{ARR}_A$$

The advantage of this method, unlike the payback period method, is that it does recognize all the cash flows over the whole life of the project. However, its general result could be misleading as it could 'hide' important details such as the fact that income flows could be highly variable, or concentrated over a particular time period. Please note that the percentage figures in this example are unrealistically high, again they are exaggerated so as to emphasize the points raised by the explanatory example.

A conclusion on conventional techniques of investment appraisal

Both of the techniques touched upon here have the great advantage of being quick and easy to use, and simple to understand. It is because of these attributes that they are frequently used, especially concerning investments in buildings at the micro level. However, the overriding criticism of them as methods of investment appraisal is that they do not specifically take into account the fact that money has a 'time value'. The **time value of money** is an important concept that highlights the ability of money to earn *interest* over time. This ability of money to earn interest is an important characteristic as it implies that money received today, in the present, is worth more to you than the *same* amount received in the future, simply because you can invest any monies received now and earn additional sums of money over time based upon the initial sum. The easiest way of appreciating the value of interest payments is to explain what happens to your money when you open a deposit account at the bank or building society. With a deposit account you will earn interest calculated by using the 'compound interest formula':

$$A = P(1 + r/100)^n$$

where: A = the sum arising if 'P' is invested in the account;
P = the initial deposit;
r = the rate of interest;
n = the number of years that you hold the account for.

For example, if you were to invest £10,000 over a twenty-five year period, at an interest rate of 10 per cent, your investment would 'grow' to a value of £108,347.06 as seen by the following worked example of the equation:

$$A = £10,000(1 + 10/100)25 = £108,347.06$$

(Remember, that the actual purchasing power of your final payout is likely to be seriously eroded by inflation.)

In other words, if money is not received until some future date, it has an *opportunity cost* attached to it. This opportunity cost is the interest payments foregone by not receiving the money today and therefore not being able to invest it. The longer one has to wait for money the greater becomes this opportunity cost of distant returns, and therefore the lesser the value that you will place upon them. Thus, one needs to devise a technique that enables the investor to appreciate that monies received now, or at early points in the investment, are actually worth more than the same nominal amounts that could be received from the investment in the future. One can achieve this by converting all future income flows into their **present value** (PV). This is especially important when examining investment in buildings due to the long time horizons involved. Such a conversion of future values to present values is achieved by using discounting techniques of investment appraisal as seen directly below. Before examining such discounting techniques, it must be stressed that the time value of money represents its ability to earn interest, and is *not* about how inflation can reduce the value of money over time. The impact of inflation would have been taken into account at an earlier stage of the investment appraisal. For example, the assessor who drew up the likely returns from the buildings in the table above would have converted nominal figures into *real* figures.

(B) Discounting techniques of investment appraisal

Discounting techniques of investment appraisal are considered to be superior to conventional techniques of investment appraisal due to the former method's implicit recognition of the time value of money (discussed immediately above). As this recognition of the time value of money is an important consideration for the investor, and because of the time scale involved with the majority of investment in buildings, discounting techniques are the preferred form of investment appraisal in relation to the built environment, especially at the macro level. Although discounting techniques are slightly more complicated than the conventional techniques, they are, with the assistance of a pocket financial calculator, still very easy to use and understand. The two main discounting techniques are: the **net present value** (NPV) technique, and the **internal rate of return** (IRR). We now go on to briefly examine these methods of investment appraisal, although the former is discussed in more detail as the latter is merely a derivative of the first.

(1) The net present value technique

This method of investment appraisal takes into account that the income received from a project is received over a number of years, and therefore it converts (discounts) these future income streams into today's value. In other words future income streams are 'deflated' in order to reflect their value to the investor at present, that is, their present value. Therefore, the net present value (NPV) is the sum of all future income streams converted into present value terms minus the initial cost of the investment. In terms of a formula this can be expressed as:

$$\mathrm{NPV} = \sum \frac{\mathrm{X}}{(1 + r/100)n} - K$$

where: X = net annual income streams;
r = the rate of interest that reflects the true opportunity cost of money;
n = the life of the project;
K = the initial capital cost of the investment.
Σ = the sum of

For example, if a project in question returned a net income flow over three years before stopping, one could work out the net present value of the project as follows, where the answer to the formula tells us what the project is worth to the investor in present day terms.

$$\text{NPV} = \sum \left[\frac{X}{(1+r/100)^1} + \frac{X}{(1+r/100)^2} + \frac{X}{(1+r/100)^3} \right] - K$$

Whereby the superscript number after each bracket denotes the applicable year for each part of the calculation. The decision rules using this technique are that a project will be worthwhile as long as the net present value is greater than zero:

$$\text{NPV} > 0$$

and that in the likely case of limited funds being made available to the investor, one would undertake the project, or projects, that promised the highest net present values. For example, assuming an opportunity cost of capital of ten per cent, one could work out the net present value of building *A* in the following way:

$$\text{NPV}_A = \sum \left[\frac{1500}{(1.1)^1} + \frac{1250}{(1.1)^2} + \frac{250}{(1.1)^3} + \frac{250}{(1.1)^4} \right] - 2500$$

$$\text{NPV}_A = [1363.64 + 1033.06 + 187.83 + 170.75] - 2500 = \underline{255.58}$$

Therefore, according to our first decision rule this project is viable as the net present value is greater than zero, so that even in present value terms a profit is likely to be made from the investment. Note that although, in nominal terms, the same amount (£250) is likely to be received in both years three and four of the project, the formula has placed a lower value on the money to be received in four years time than the money to be received in three years time. As detailed above, this lesser present value given to the figure in the fourth year reflects the opportunity cost of having to wait for your money and therefore lost ability to invest it. Effectively this formula is the opposite of the compound interest formula. For example, if you had £170.75 and invested it over four years at an interest rate of ten per cent it would accumulate to a total value of £250.00. If one were to calculate the net present value for building *B* in exactly the same manner as above one would arrive at a net present value figure of:

$$\text{NPV}_B = \underline{669.86}$$

As this answer is greater than that obtained for building *A*, building *B* would represent a better investment than building *A*, and therefore would be the preferred choice in the light of limited funds. This preference is normally expressed in the following way:

$$\text{NPV}_B > \text{NPV}_A$$

It should be noted that to increase the level of caution in one's results a **risk premium** can be added. Such a risk premium, say two per cent, would be added to the value for r that was initially chosen, and this will have the effect of devaluing future income flows yet further, and therefore emphasizing any marginal projects. The more pessimistic, or risk averse, investors are, they are likely to select a higher risk premium than the one chosen by the optimist. To clarify this point try repeating the above calculations with a value of r equal to twelve per cent for example. You will quickly see that by incorporating a higher value of r into your equation it has reduced the perceived net present values of both buildings, thus making them look less attractive as investments.

A problem though with the process of discounting, is that it is biased against projects that yield higher returns in the future, as was also the case with the simple conventional payback procedure. Another common criticism with this method of investment appraisal is that the investor has to select a value of r that accurately reflects the opportunity cost of capital. Where this cannot be done one could utilize the alternative discounting technique, namely: the internal rate of return. This technique will now be briefly discussed.

(2) The internal rate of return

The internal rate of return (IRR) is merely an extension of the net present value technique, whereby the IRR itself is simply the value of r that would set the net present value calculation to zero. Therefore, if a project, such as the construction and letting of building *A*, were to seem viable at a value of r equal to 10 per cent, this may not be the case if we had used a higher value of r, say of 15 per cent. By calculating different values for the NPV of a project one could draw an IRR schedule as shown in Fig. 13.1 which shows at which value of r our NPV calculation would be equal to zero. The relevance of this figure, say 16.5 per cent, is that as long as the opportunity cost of capital were less than the IRR the project would be profitable, and if it were more, losses would be made. It therefore gives us a 'clear' dividing line between project success and project failure. This line may not be as

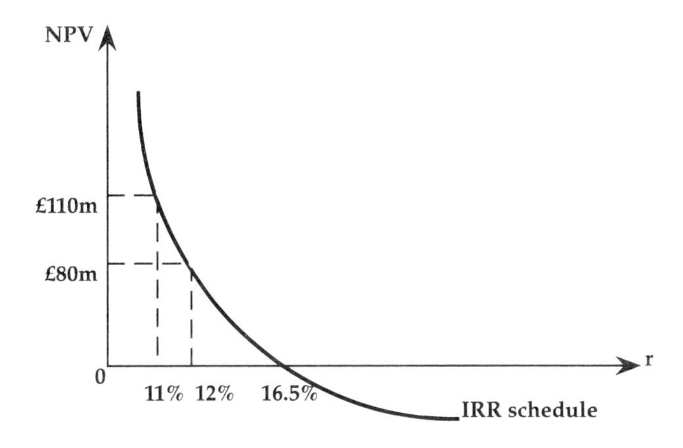

Figure 13.1. *The internal rate of return*

exact as that implied by the diagram simply because future values are, by definition, uncertain. Moreover, the calculation of the IRR is complex and one would be best advised to use a good financial calculator for this purpose.

A conclusion on discounting techniques of investment appraisal

As already stated, discounting techniques are felt to be more accurate inasmuch as they accommodate the principle of the time value of money. However, due to their relative complexity they are not always used at the micro level, nor are they necessarily appropriate when examining very short run investment situations.

A conclusion on investment appraisal techniques

It can be seen that different techniques of investment appraisal can recommend different projects to the investor. Using the hypothetical case studies of two buildings from the table above, one finds that the payback period method recommends investing in building *A*, whereas the average rate of return method, and the net present value method, suggest that building *B* represents the best investment potential. This apparent conflict of results does *not* mean that any one of these techniques are incorrect, it simply highlights the fact that they have been designed to meet the requirements of different investors. For example, some investors may need to take a short-run view of an investment, whereas others may have the luxury of taking a longer term view. Another issue that must be fully understood when interpreting the results of such calculations is that they are only as good as the data that is fed in to them. The old adage: 'garbage in, garbage out' – certainly holds here just as it does with any use of figures. Problems of the accuracy of the data used can arise for a variety of reasons as below for example.

1 Future cash flows are estimates, and even the best thought out forecasts can be rendered inaccurate by the advent of unforeseen circumstances. A war in one of the main oil producing areas of the world could push up energy prices. Such an escalation in energy prices can not only lead to an increase in the operating costs of a building in terms of heat and light, but they can also 'trigger off' an economic recession that depresses the demand, and therefore, the rental growth prospects of such buildings.

2 The techniques assume that we know the life of the project. Again the longer one tries to predict this figure into the future, the less certain it becomes. A new technological revolution could make some buildings become functionally obsolete although it was originally thought that they would have quite long economic lives.

3 Discounting techniques assume a constant value of *r* throughout the calculation, whereas in reality the opportunity cost of capital to the firm is likely to change and fluctuate quite significantly over time depending upon alternative investment possibilities.

4 One may assume that one figure that would be certain is the initial development cost of the project. However, this too can diverge from the original forecasts quite considerably. Development costs for the buildings may increase because a rise in the interest rate filters through to higher cost development finance, or an unanticipated shortage in some building materials pushes up construction costs. To illustrate this point try to find some examples of the cost of a completed building as compared with the contractor's initial tender. In the case of large buildings and civil engineering structures this difference is usually so marked that it often becomes difficult to see the relationship between the two.

Despite these potential shortcomings forecasting of input data can hopefully reduce them, thus leaving us with sensible investment appraisal techniques and subsequent analysis. One can now take these principles yet further, and increase the level of application, by examining how one could determine the economic life of a building, or a group of buildings.

The determination of the 'economic life' of a building

Obviously the physical life of a building can be very long, with many buildings remaining structurally intact, and actually in use for hundreds of years. However, in economic terms it may be found that such buildings have exceeded their useful (profit maximizing) life only after a relatively short period of time. If a new production process is devised in a technological breakthrough, existing, older factories, may become obsolete, and as such the 'economic life' of these buildings will have come to an end. In fact, even a change in a nearby land use from one type of building to another can result in either the lengthening of the economic life of the building in question, or the hastening of its economic demise, depending upon whether the new buildings are of a complementary or conflicting nature. Therefore, it would seem apparent that we need a logical framework to enable us to make well-informed judgements about the economic life of a building, and therefore enable entrepreneurs to assess the potential for the redevelopment of their property. However, throughout this analysis it must be recognized that the decision whether or not to knock down a building and redevelop the site, will be constrained by a number of important factors such as planning regulations, historical preservation orders, public pressure groups and the consideration of alternative solutions to the problem. In relation to the last point, it should be realized at the outset that if a building has reached the end of its economic life it need not necessarily have to be demolished and replaced in order to make a profit. Alternatively, the economic life of the existing building could be extended by extensive refurbishment, or a change in use.

A variety of investment 'models' are available in order to assist one in determining the economic life of a building. Essentially they all put forward the recommendation that:

A property should undergo change when the present value of expected future net returns from the existing building, or use of land, becomes less than, or equal to, the net value of the next best use that the site could be put to, after taking all development costs into consideration.

The changes that actually occur could range from redevelopment to refurbishment and this decision would depend upon a variety of factors such as finance and market demand. For example, the owners of the site may not have sufficient funds to redevelop the site, or alternatively market research may reveal that some clients view it as a fashionable asset to be seen occupying a converted traditional building rather than a modern one. In assessing such a calculation the concept of present value is used as the life of buildings is normally long and as such there is a need to recognize the time value of money (see p. 214). 'Real figures' should also be used otherwise the erosive power of inflation can distort the accuracy and usefulness of one's calculations. It is also important to note that this model is not static, but dynamic. That is, it would be naïve to work out a result today in the hope that your answers or predictions would be fully accurate in twenty years time. Variables within the model do, and will, change over time, and therefore the model needs to be frequently recalculated in order to check the accuracy of its initial answer. Such a model could easily be run on a computer program so as to facilitate the task here, as one would merely have to feed in information regarding the values of changing variables. For example, if one ran the model today for a particular building with both known and forecasted values for your variables, the model may suggest that the building in question, in its present form, is likely to reach the end of its economic life in say thirty years time. However, circumstances can obviously change – the building may suffer from previously unexpected high repair bills because of a structural fault, or nearby infrastructure may be improved at a later date making the building, or its site, more accessible and therefore more attractive. In the event of such changes one would have to recalculate the model, with the redefined parameters, to see whether the 'terminal date' had been altered or not. Generally, the original accuracy of your model will depend primarily upon the accuracy of your original forecast.

In order to ascertain the present value of the expected future net returns to be gained from the *existing building*, or use of land, one needs to initially calculate values for the buildings anticipated **net annual returns**. Net annual returns are simply all the monies received from the building such as rents (**gross annual returns**) *minus* the **operating costs** of that building such as the costs of ongoing repair and maintenance. That is, in terms of a simple expression, one can write:

$$NAR = GAR - OCs'$$

where: NAR = net annual returns;
GAR = gross annual returns;
OCs' = operating costs.

In order to accommodate the process of discounting one can rewrite this expression as:

$$NAR = \sum \frac{GAR - OCs'}{(1 + r)^n}$$

Remember that the figures obtained for both gross annual returns and operating costs must be estimates as they occur in the future.

As said above, gross annual returns are the rental received from the leasing of a building, and such rental is largely determined by the level of demand for the building in relation to other available supply that the tenants may choose from. The gross annual returns of a building are likely to decrease over time, at least in real terms, because of a number of factors:

1 When the building was new it may have been perceived as being unique and 'state of the art', and therefore attracted a high demand and subsequently a high rental. However, the success of such a building is likely to encourage similar developments as other entrepreneurs see that good profits are to be gained by the letting of such buildings. Therefore, as a consequence of this, new and similar buildings are built which will compete for occupants with the existing one. This resultant competition increases the total supply of floor space of the type of building in question, and if demand does not rise as fast as supply, rents will be eventually driven down. Note that changes in rent may not be immediate as alterations to rents may only be possible during periodic, rather than continuous, rent reviews.

2 As the building ages, higher repair bills could decrease rent levels over time as tenants refuse to pay high rents for a building that is constantly undergoing repairs, or is in need of attention, such as the lift continually being out of order.

3 Tenants may be less willing to pay a high rent for a building that is ageing for a variety of reasons other than that discussed in point (2) directly above. The building may become technically obsolete, or the area in which it is located may no longer be suitable for its effective operation. That is, the land and other buildings in the vicinity may become run down, or suppliers may have moved on, the transport infrastructure may have changed, and so on.

4 As there is increasing risk and uncertainty as one tries to forecast further into the future, valuations of gross annual returns tend to become increasingly more cautious and therefore conservative, the further into the life of the building one tries to predict.

5 The process of discounting future values into present value terms will cause an automatic tendency for future returns to decline over time.

As gross annual returns tend to decline in real terms as the building ages, its operating costs are likely to escalate. Operating costs increase over time for a variety of reasons:

1 The structure and fabric of the building deteriorates physically, and is thus subject to increasing repair and maintenance.

2 Older buildings are difficult and costly to adapt to new technical requirements and demands, such as a change in use, or the need to incorporate improved energy savings devices such as double glazing.

3 Changes in legislation concerning health and safety, fire prevention, or energy wastage for example, could all require monies being spent on older buildings in order to bring them in line with current, acceptable standards. For example: the removal from old buildings of asbestos, or other materials that are now recognized as being dangerous to the users of the building.

As net annual returns are simply the difference between gross annual returns and operating costs, one will see that the net annual returns of a building will decline as it ages as gross annual returns fall and operating costs rise. In order to determine the economic life of a building, however, one does not just wait until the net annual returns of the existing building fall to zero. If one were to behave in this manner, the built environment would contain a large number of poor quality buildings and urban land would not be put to its most profitable use. So as not to miss out on profitable development opportunities, one also needs to periodically assess the present value of the *next best use* that the site can be put to, after taking redevelopment costs into account. One could express the estimate of the value of the land in its next best use in equation form:

$$\text{The value of the next best use} = \sum \frac{GAR^* - OCs^*}{(1 + r)^n} - (D + C + B)$$

where:
- GAR^* = gross annual returns from next best use (building);
- OCs^* = operating costs of the new building;
- D = demolition costs of existing structure;
- C = site clearance costs;
- B = rebuilding costs.

When the original building was constructed the expected value of the next best use must have been less than the anticipated value of the building that was built, otherwise the alternative building would have been erected instead of the one that currently stands. However, the value of the next best use that the site can be put to is likely to rise as one is able to build a building that reflects the current state of the economy and technology. The demands for such a 'state of the art' building are likely to be high as was the case of the original building when it first appeared on the market at the beginning of its economic life. Therefore, one will reach the stage whereby:

The value of expected future net returns from the existing building, or use of land, equals, or becomes less than, the anticipated net value of the site's next best use after taking all costs into consideration.

Once one approaches this situation a variety of decisions about the existing building, or land use, should be made. Essentially one could either:

1 demolish the existing building and replace it with another, more profitable, one; or:

2 extend the economic life of the existing building by:
(i) Extensive refurbishment so as to attract higher rents for the improved facilities and working environment;
(ii) Create a change in use in order to find a client who is willing to go on using the building;
(iii) Reduce the operating costs of the building, and thus increase net annual returns (remember that NAR = GAR – OCs'). Reductions in operating costs could be achieved by investing in improved energy efficiency within the building so as to reduce energy costs for example.

In conclusion, it must be said that this 'model' is certainly a useful decision aid when determining the economic life of a building. However, as is the case with all models, it is unlikely ever to be 100 per cent accurate. Any model is only as good as the quality of the information fed into it in the first instance, but, as importantly, the magnitude of key variables in the model is likely to change over time. This latter point is of particular importance when looking at investment in buildings as their economic life may span several decades if not more. Thus, it is again stressed that the model must be viewed as being dynamic and not static. Obviously the model is also continuous as any new building built, or any change to the existing building, will have a limited life in both economic and physical terms. In other words, the model should indicate a progression of development on a site as buildings are replaced by others as time progresses. We will now go on to consider some examples of how this model can be used in a dynamic way, and indicate how factors external to the model (exogenous variables) can alter the economic life of a building.

Factors influencing the economic life of a building, or a group of buildings

There now follows a variety of instances by which the economic life of a particular building, or indeed a group of buildings, can change over time in response to a number of influences in the urban market. It is important to appreciate, as seen in Chapter 12, that the evolution of any urban area is a continual one which will obviously have implications for any building within that area. For purposes of illustration I shall firstly examine the case of industrial buildings.

(A) Industrial buildings in the inner urban area

Because of the process of urbanization, and the importance of industrialization in that process, much of the inner urban area can be characterized by the dominance of old, industrial buildings. Such buildings were often the result of the 'industrial revolution', but their specific location was also derived from a smaller urban area, and a different transport structure to the one that exists today. Thus, for example, many nineteenth-century factories were built around previous rail junctions that are now no longer in existence. Therefore, it is easy to realize that developments over time have led to the shortening of the economic life of such buildings although they are still often physically sound. The evidence to demonstrate this point is easy to see as old industrial buildings are often left derelict, others are replaced by offices, or alternative land uses such as car parks. As seen in Chapter 12, one of the main forces at work here was the process of 'decentralization'. The forces of decentralization created the situation whereby many manufacturing firms preferred to locate on the periphery of a town rather than near its centre. For example, in many countries rail freight no longer goes to the heart of the urban area and as such industry relies on supply and distribution by road. Thus, firms needed to locate near the major roadways and did not wish to be hampered by the congestion so often associated with the inner urban area. Moreover, modern production techniques favour large single storey industrial units (the advent of the fork-lift truck was an important development here), rather than the multi - storey lay out of pre-twentieth century buildings. Such modern units therefore require a large amount of land, and thus siting near the centre of a town would be cost prohibitive due to higher land prices there. Therefore, the demand for these older, centrally located buildings as prime manufacturing property declined. As a result, the gross annual returns received from such buildings declined which in turn reduced their net annual returns. Moreover, the operating costs of older buildings can suddenly rise as unanticipated repairs become more frequent, or legislation – for example such as improved health and safety legislation – forces expenditure on the building – such as enhanced fire precautions. This again will lower net annual returns as an increasing proportion of gross annual returns are eroded by cost increases. In other words, the economic life of old, central, industrial buildings has been shortened by developments such as the growth of a town, changes in technology, and changes in transport.

(B) Inner city housing

Inner city housing is now used as a vehicle to demonstrate how the model can be applied to examine the economic life of residential buildings. As urban areas develop it is often found that a large stock of housing remains within the inner city, although it has often been observed that the main driving force for many years has been the decentralization of houses towards the more spacious and affluent suburbs, (see Chapter 12). Such remaining inner city housing is often of poor quality and is in competition with other land uses. Thus, the question that has to be asked is: 'Why do these houses still exist as they must surely have reached the end of their economic lives?' The reasons that such housing is still in existence can be explained in a number of ways:

1 Houses are often kept for 'consumption' rather than financial reasons, and as such, because of the non-economic motive in some cases, the model is largely inapplicable. For example, elderly people may not wish to move out of a community in which they have lived all of their lives despite the fact that there may be superior alternative accommodation elsewhere.

2 Planners may wish to promote, or keep, a mixed land use in the central area, with housing as an important contributory factor. As such, local authorities may attempt to enhance the economic life of centrally placed residential buildings by providing improvement grants for instance.

3 In pure economic terms, the economic life of such houses has been extended by the private sector by:
(i) Landlords increasing the gross annual returns received from such buildings by subdividing large properties into smaller units and thus letting the building to a larger number of people. Although the rental per person is likely to be reduced in such circumstances the total receipts from *all* occupants are likely to be greater than would be the case if the building were to be leased to a single, more affluent client. In many developed nations there are numerous examples of three storey nineteenth-century houses being subdivided into several flats.
(ii) Landlords have maintained net annual returns by keeping operating costs to the barest minimum. The evidence of such behaviour can be seen by the poor, dilapidated appearance of many rental properties.

Therefore, we can see a variety of circumstances that will lead to increases in net annual returns which can thus prolong the economic life of a building.

(C) Changes in urban land use

The use of adjacent or nearby land to a building, or group of buildings, will also have implications for the economic lives of those existing buildings. For example, if a complementary land use were to be put alongside, or near, the existing buildings, the demand, and hence the net annual returns of the original buildings, could increase. Imagine older factory buildings being made more appealing by the development of improved local transport infrastructure, or the building of other industrial units thus encouraging the onset of agglomeration economies. Conversely, conflicting land uses could have the opposite effect and actually reduce the net annual returns, and thus the economic life, of existing buildings. If industry was to be sited near present housing the positive effects would be increased employment, but on the negative side few people may demand such properties due to the high levels of pollution and congestion created by such industry. Therefore, the economic life of such adjacent residential properties could be reduced.

(D) Developments in construction technology

Changes (improvements) in construction technology could hasten the end of the economic life of a building as the redevelopment of the site becomes an increasingly attractive (profitable) proposition. Changes such as improved technology in the area of prefabrication, for example, could enhance construction productivity and thus enable the owner of the land to replace any existing building both at lower cost and more speedily than would otherwise be the case, and therefore enable them to redevelop the site at an earlier date.

(E) Government activity

The activities of the public sector can alter the economic life of a building, or group of buildings, in a number of ways. For example:

1. Although a building may have reached the end of its economic life, planning restrictions may prevent the site's redevelopment. Such instances are common in the case of buildings which are protected because of their specific historical interest (see below).
2. Compulsory purchase orders (CPOs) can be used by local authorities to speed up redevelopment in one instance, or extend the economic life of other buildings in a different situation. An example of the former could simply be a CPO being served on a building to make way for a new roadway. In such an instance the economic life and physical life of the building are prematurely ended. In the latter case CPOs could be used to 'zone out' competing land uses and replace them with complementary land uses thus enhancing the economic lives of existing buildings.
3. Improvement grants can enhance the economic life of a building by enhancing its attractiveness to potential occupants, thus enabling higher rents to be charged which would lead to increased net annual returns.
4. Interest rate policy can also have an important bearing on the economic life of a building as the cost of development finance is directly linked to the interest rate. Thus, for example, if the government were to lower the interest rate, as part of an expansionary demand management policy, (see Chapter 11) this could lower the costs of redevelopment and thus encourage early new build.

The case of historical buildings

Increasingly people are becoming aware of the need to preserve examples of architecture from previous ages and to conserve buildings of particular historical or national interest. Using the model (above) in its 'raw' (profit maximizing) form is obviously not appropriate here as the recommendations of the model do not take into account the external benefits derived from buildings of historical interest such as the pleasure people gain from visiting or simply seeing such buildings. Therefore, as the market ignores social costs and benefits there is a rationale for governments to intervene to preserve such buildings. This could be done in a number of ways:

1 Bring the building under *public ownership*, (perhaps convert it into a museum).

2 Leave the building under private ownership but provide a *subsidy* to increase net annual returns received by its owners.

3 Buildings can be *listed* and once listed they cannot be tampered with without strict approval let alone demolished.

4 One could allow, within acceptable limits, a *conversion* to a more profitable usage. Old houses can be turned into office accommodation without greatly changing the external appearance of such buildings. In fact some developments have maintained the original façade of a building, yet completely rebuilt the structure behind it.

A conclusion on private investment and the built environment

It can be seen that the decisions of the private sector, with profitability as its primary motive, can create substantial changes to the built environment in terms of the location of buildings, their use, and the rate at which they are changed. However, as seen from the analysis above, many influences on the built environment are beyond the control of the private investor, moreover there are also a range of public sector constraints imposed on any private investment decision.

14

Public sector investment appraisal and the built environment

The role of the government sector in the direct provision and maintenance of buildings, and the degree to which the state gets involved in the markets of the built environment such as the housing market, varies from country to country. However, the public sector is normally concerned with a wide range of building programmes in the urban area ranging from the construction of hospitals, and schools to the building of infrastructure such as motorways and sewage treatment plants for example. Just as with the private sector, the government will wish to maximize its returns from 'investments' into building projects. However, whereas the private sector are likely to be mainly concerned with maximizing **profits** from a building, the public sector are likely to be concerned with other criteria such as the maximization of **social welfare**. In order to cater for this social goal the standard investment appraisal techniques examined in Chapter 13 are no longer appropriate in their existing form, and therefore either need to be adjusted to take on a new set of criteria, or abandoned completely in favour of investment appraisal techniques specifically designed for the public sector. It must be said though that the distinction between investment appraisal in the two sectors is not always very clear. The reasoning behind this statement is due to two recent developments:

1 In many countries the public sector has been encouraged to become more cost effective and accountable, and in such situations one is likely to see the use of standard private sector investment appraisal techniques and principles. This has come about as governments have required many parts of the public sector to become self-funding, or at least to strive for that goal. Moreover, the voting public seem increasingly aware of, and concerned about, any perceived wastage of tax payers' monies.

2 The private sector has realized that it can actually enhance its profits and image by being more aware of public needs. For example the total development proceeds of an intensively developed site may well be less than one which incorporates adequate open public space and good quality landscaping. The latter development is likely to be more socially desirable and could thus command a higher overall sales price.

(A) Private costs and benefits

These are the costs and benefits which directly affect the private investor. If the entrepreneur is considering whether or not to invest in a project by setting up a factory, for example, his or her decision is frequently determined by the profit motive. Thus, the underlying question in such an instance is: 'Is the investment a profitable one?' Therefore, the investor is concerned with the fixed costs and the variable costs of the exercise (total costs), as well as the revenues that can be expected from the project. These are explicit costs and returns (benefits), which the firm must take into account. These private costs and benefits are also sometimes termed 'internal costs and returns' as the firm includes them in their internal accounts when making a decision on an investment (see Chapter 5). However, there may be some costs and benefits that result from an entrepreneur's actions which are imposed upon others, of which the private entrepreneur will take no account. That is, there are what are termed 'spill over' costs and benefits caused by an investment. These are often termed 'social costs' and 'social benefits'.

(B) Social costs and benefits

It is felt by most that the government ought to undertake social investment appraisal so as to attempt to maximize the net social benefit (NSB) of society from public projects. Profit maximization is seen as a dangerous motive for government as it would be quite easy for government to achieve this objective in many instances as many public concerns are monopolistic in nature. To achieve this goal of maximizing net social benefit the public sector project appraiser needs to identify social costs and social benefits derived from the project in question. Such costs and benefits are simply those which effect society as a whole. Imagine if a factory were to be built in a particular area as part of an area regeneration programme: on the negative side, existing local residents may suffer from increased levels of air pollution created by the industry, increased noise (from machinery and transport), increased traffic congestion, and, moreover, the 'scenic virtues' of an area may be adversely affected. That is, negative externalities exist – costs which are external to the manufacturer. When such external costs are added to private costs we have the true social costs of the action. In a similar way, external benefits may arise from an investment. The building of a factory could lead to new jobs being found for workers in the area who would otherwise be unemployed. Such a benefit could be compounded by the 'multiplier process'. Thus, there is a social benefit to the local community.

A technique of investment appraisal that is available in order to attempt to assess such social costs and benefits is **cost-benefit analysis** (CBA). This technique is designed to give guidance to the decision maker in order to be able to judge the 'worthwhileness' of proposed public sector investment projects, especially with respect to the provision of 'public goods' such as open public space in the urban area (simply because it is these goods that private sector techniques are likely to under provide). Moreover, this technique should bring about a more efficient allocation of resources, and maximize net social benefit where:

$$\text{NSB} = \text{Social benefits} - \text{Social costs}$$

Because of a variety of forms of market failure (see Chapter 15) the free market system would only maximize NSB under extremely restrictive assumptions – namely the assumptions of 'perfect competition' (see Chapter 7). Such a situation of perfect competition is unlikely in reality, and moreover, most public sector outputs are not allocated via markets. Furthermore, as suggested above, the market fails to produce sufficient quantities of some goods (public goods and merit goods), and too many of those goods creating 'negative externalities'. Furthermore, CBA is a technique which seeks to bring greater objectivity into public sector decision making, as before the advent of CBA many more public projects were based upon the subjective political whims of members of the government and civil service rather than logical thinking. Today, one frequently finds that public sector projects involve conducting a CBA study in order to ascertain the overall benefits to society of constructing roads, bridges, airports, parks and new urban areas and the preservation of existing buildings of historic interest.

(C) The use of cost benefit analysis

Using this technique a public sector project such as an urban redevelopment programme, or the construction of a new road, will be considered to be viable as long as the 'social benefits' outweigh the 'social costs' incurred. When considering a project under CBA one must follow an agreed standardized procedure for purposes of consistency and comparison.

The CBA procedure – 'the CBA balance sheet'

Stage 1 The listing of all the relevant costs and benefits attributable to the project.

Stage 2 All costs and benefits are then evaluated so that they can be expressed as a common monetary value.

Stage 3 All costs and benefits are then discounted back to the present.

Stage 4 A degree of risk and uncertainty will exist in all projects, and the degree to which this affects the project needs to be identified.

Stage 5 Making allowances for constraints on the project. Even though a project may appear desirable there may exist constraints, such as political pressure or financial shortfalls, that may force the cancellation of the project or the acceptance of a 'second best'.

Stage 6 Project appraisal. Finally, the decision maker needs to assess all the costs and benefits, and select a project that yields the best increase in 'social welfare', or net social benefit (NSB).

Thus, for each proposed project we need to run through a procedure such as this. However, as we shall now see, the completion of each step is fraught with difficulties.

The CBA procedure: problems encountered

Problem 1 The choice of costs and benefits to include: ideally a CBA calculation should take into account all the costs and benefits incurred by the project. However, in practice, only the immediately relevant effects are taken into account. Thus, one has to decide upon a 'cut off' line. For example with the construction of a new motorway, major benefits for the project, such as reduced time taken for journeys, or the reduction in congestion on existing roads would be taken into account. But, a factor such as the deterioration of the general landscape *may* be considered too insignificant, or too difficult to calculate, to allow for it in the analysis. That is, the decision maker has to draw the line somewhere by deciding which costs and benefits are immediately relevant and which are not.

Problem 2 The evaluation of the selected costs and benefits. Having decided upon what costs and benefits to include, the choice of values to attribute to them is an obvious problem. That is, it is necessary that we convert the 'value' of all items in our analysis into a common comparable dimension, and as most of us are dealing in financial units analysts attempt to convert these items so that they all have *monetary values*. When undertaking this exercise two possible situations can confront us.

1 Some items in our CBA will have market prices, and therefore when we can use these there is no problem. Although, in many cases one may have to adjust these market prices if they do not properly reflect the true social costs and benefits involved.

2 Some items have no price at all, and therefore some alternative form of evaluation will have to be devised.

Therefore we need some technique to either assign monetary values in the first place or adjust existing market prices so that they reflect social valuations. The technique devised to attempt to solve this problem is known as **Shadow Pricing**. A shadow price is the price determined by the project evaluator in order to reflect the true social costs or value of a good or service. As implied above, when assigning shadow prices one can be faced with two possible situations.

1 One could have certain items in the CBA where the market price of the item does not reflect the true social costs of that item. A common example is when we have the situation whereby the labour used on a project may

otherwise be unemployed if the project did not exist. Thus, the 'opportunity cost' associated with the employment of these people is zero, and thus it is thought that to value the labour at the market wage would be incorrect as it is costing society nothing in terms of alternative output forgone. Similarly, using the same logic, the social cost of using idle, uncultivated land of no scenic interest, is zero. However, if the land was in short supply, and had alternative uses, output forgone must be costed.

2 There may be many costs and benefits in our project which have no market price at all. The imputation of monetary values in such cases can be difficult and time consuming. Depending upon the problem encountered we can use one of two approaches in order to establish shadow prices for them: (i) the inference approach; (ii) the questionnaire approach.
(i) The 'Inference Approach' is concerned with the fact that in some cases it may be possible to adapt prices found in similar situations where market prices already exist. Thus, we estimate our value based upon other similar values. For example the benefits to local residents of a new public golf course providing free facilities to the community could be gauged by observing the prices which people are prepared to pay to use a similar facility provided privately elsewhere. That is, we use the concept of **willingness to pay** (WTP) in order to assign values to items in the study. However, with many collective goods no alternative market price exists at all and thus one can use the 'questionnaire technique':
(ii) The 'Questionnaire Technique' is where one would ask the public to consider what value they place on various costs and benefits. That is, we wish to ascertain their willingness to pay for a particular project. On finding this, we could then aggregate people's 'willingness to pay' to form the 'market demand' for the project.

Problem 3 Problems in discounting. Obviously any project undertaken will produce a stream of costs and benefits that occur over a number of years, and therefore to find out the total value of the stream of net benefits it is necessary to convert the net benefit in each time period to its present value via discounting. When confronted by a number of investment projects, in order to maximize utility, the decision maker should select the projects that yield the highest returns in terms of present value – provided that the project is affordable in the first instance, and will not be affected by other constraints. Thus, the public sector decision maker has to work out which project yields the highest benefits in terms of the present value of Net Social Benefits by using the following formula.

$$NSB = \sum \frac{SB - SC}{(1+r)n} - K$$

where SB = social benefit;
SC = social cost;
n = the anticipated life of the project;
r = the selected discount rate;
K = the development costs of the project in social terms.

However, there are problems here in assigning the correct discount rate to the formula. If we postpone current consumption, resources can be used for investment in capital goods which will produce greater output in the future. That is there will be growth. However, because future consumption is valued less highly, by most, than present consumption, societies time preference tends to lead to a preference for present consumption. Present consumption is preferred because:

(i) people tend to suffer from 'myopia' – they cannot, or are unwilling, to look into the future;
(ii) future income and other benefits are less certain because few can claim to accurately foresee future events;
(iii) future income is likely to be greater than present income, and thus the marginal utility of present income is higher than that of future income.

Therefore, left to the free market one is likely to arrive at short-run consumption and investment behaviour that may not adequately take the interests of future generations into account. Thus, when considering public projects, one should really consider: the **social rate of time preference** as this recognizes that:

1 Society's time preference is not the same as the sum of 'individual time preferences'. This is because individuals suffer from myopia inasmuch as it is felt that they would fail to see how much they would benefit from future consumption as opposed to present consumption. Moreover, many may like to save more, but will not do so unless they are sure that others will do so also.

2 More importantly, the government is responsible for the welfare of future generations and not just the people of today. Thus, investment must also take the future into account.

Both of these arguments give justification for discounting public projects at a lower rate of interest than arrived at through the market. In practice, governments have set a (lower) rate of interest known as the 'test discount rate,' which is set lower than the market rate in order to assist the public sector decision maker.

Problem 4 Risk and uncertainty. Just as a business faces uncertainty there will also be uncertainty concerning public programmes. Although risk and uncertainty are conceptually two different things, in most CBA writings the difference between the two is ignored, and they are both normally considered under the heading of 'risk'. However, I shall briefly consider the distinction between the two.

1 Risk. A risky situation is one in which we are not sure of the outcomes of a project, but is where one can assign probability values to the likely outcomes. In this situation one could draw a probability distribution representing these likely outcomes. Such probabilities may be assigned objectively by using prior knowledge from similar existing projects, or

subjectively being based on the decision maker's own feelings, forecasts and calculations.

2 Uncertainty. This refers to the situation where the likelihood of different outcomes is completely unknown. However, such situations are usually subjectively turned into risk situations.

Although some see the problem of risk and uncertainty as one worthy of extensive consideration, others have suggested that for public sector investment decisions risk is not necessarily an obstacle because it is relatively insignificant. The logic behind this view is that not only do governments usually have vast financial reserves, they undertake so many projects that it can be said to have a 'diversified' portfolio, and the losses incurred are usually offset by gains. This is known as **Risk Pooling**, as the degree of risk associated with each project is spread over a large number of other projects. Moreover, experience would hopefully lead to a situation where successes outweighed failures. The failed projects can then be simply counted as costs covered by the successful projects. Thus, the government does not have to adopt the 'Safety First' approach. In order to ensure a satisfactory diversification of risk however, the following two rules have to be adhered to:

Rule 1 No project should be so large that it dominates the investment portfolio

Rule 2 Not too many investments should be inter related. That is, if one fails this should not have a 'knock-on effect' that would damage other projects.

Problem 5 Making allowances for constraints on the project. Before embarking upon a project, one must be aware that there are several potential constraints in the way.

1 Political Constraints. Most projects involve much political wrangling. For example, in constructing a new road, not only may there be more than one local authority, or public body involved, but they may also be motivated by different interests and political views. Moreover, there may be conflicts between central and local governments.

2 Budget Constraints. When finance is in short supply we may have to rank projects according to their cost, and often projects yielding the highest NSB may have to be forgone if they are too expensive. This is especially relevant in recent times as governments have had to become more financially accountable.

3 Legal Constraints. Legal constraints may prevent a project, or part of that project being undertaken if it were to conflict with another's legal interest. It seems a notable feature of modern society that an increasing number of pressure groups are making their voice heard. For example people may object to a new development if it affects their right of light, or access, or displaces an existing community.

4 Distributional Constraints. A project is often opposed on distributional grounds. The pubic evaluator should take care to ensure that any project does not benefit the better off in society to a greater extent than those who are already comparatively disadvantaged.

Problem 6 Assessing the project's overall viability. The project with the highest net social benefit should be undertaken (subject to constraints). However, any project with a positive net social benefit would be socially desirable.

Conclusions on public sector investment appraisal

Although cost benefit analysis as a technique of investment appraisal is characterized by a large number of problems, it at least gives the public sector decision maker a degree of objectivity when examining the social desirability of proposed projects. Moreover, a better solution has yet to be devised and commonly utilized.

— 15 —

The rationale for public intervention in the built environment

Many argue that if the formation of the built environment were left entirely to the workings of the private sector (the market) many undesirable results would occur, and therefore the villages, towns and cities that we lived in would suffer from a variety of inadequacies. This standpoint stems from the belief that the market fails in a variety of circumstances in the urban environment – the term market failure is normally used to categorize these problems. This is certainly not to say that the market is rejected completely as a useful mechanism in which to shape our urban areas, it is simply stating that we must recognize that the market may give us sub-optimal results in some instances. Once these areas of market failure have been recognized it is hoped that careful public sector intervention via the use of urban economic policy could solve, or at least reduce, their adverse consequences. In fact, as one will see shortly, both **land use planning** and **building regulations**, for example can be viewed as public sector responses to market failure. However, it must also be said that public intervention is not necessarily always the correct solution to market failure as government policy itself has been known to exacerbate difficulties, or at least create undesirable 'spin offs' from the original policy. In other words, far from curing the problem, public initiatives can either fail to solve a problem, make it worse, or create others that were not there in the first instance. Thus the following text will take a critical and analytical standpoint when examining the role of urban economic policy as a means of correcting market failure in order to achieve the optimal result in the formation of our urban areas.

Several areas of potential market failure can be identified in the built environment and these are listed below before going into a more detailed examination of each in turn.

1 externalities;
2 public goods;
3 monopoly situations;
4 imperfect information;
5 societies decisions.

(1) The problem of 'externalities' in the built environment

Externalities occur when the costs *or* benefits, from a particular good or service, to society as a whole, are not adequately reflected in the market price for that good or service. It is important to realize that one can observe examples of both **negative externalities** and **positive externalities** in the built environment. Negative externalities arise because of the detrimental impact on society of certain actions by others, and positive externalities arise in situations where society benefits from the actions of such third parties. However, most texts and articles normally concentrate on the former category of externality as it is the negative aspects of living which unfortunately normally generate the most controversy and are seen as being more emotive areas of discussion. As externalities are not reflected in market prices they are said to exhibit **non-pricing**. Moreover as externalities have implications for people in society other than the direct user or users of the good or service, they are also said to be characterized by **interdependency**. It is for this reason that externalities are frequently referred to as **spill over costs** in the case of negative externalities, and **spill over benefits** in the case of positive externalities, as interdependency implies that other people are affected by the actions of others. In conclusion, externalities arise because there are a number of instances when consumers and/or producers will take decisions in the light of their own costs and benefits (internal costs and benefits) that will produce an impact on the welfare or output of others in ways that are not reflected in the prices facing those consumers or producers. It is felt that some form of public intervention should be used to address the issue of externalities as it is feared that if the market were left to its own it would create too many instances of negative externalities, and would encourage too little activity that promoted the existence of positive externalities. So as to understand this issue more clearly, we now go on to briefly analyse the cases of both negative and positive externalities using examples drawn from the built environment. After this discussion, various forms of related public intervention are examined, although the analysis does concentrate on **town planning** as the example best suited to demonstrate the relevant issues.

(a) The case of a negative externality

As seen above, a negative externality is created when the behaviour of an individual or group of individuals adversely affects the welfare of another individual or group of individuals. Such problems are not captured by the market mechanism as the 'producers' of the ill effects do not have to pay for this aspect of their actions, and those who are affected by them receive no compensation. The list of such negative externalities in the modern built environment is sadly almost an endless one, but some of the more obvious ones are listed below.

1 *Air pollution* created by industry located in or near the built environment. Note that although the industrialist obviously has to pay for the productive process itself, in the absence of public intervention in order to reduce the emission of harmful pollutants, they will not have to pay for the damage created by the spill over effect (pollution) of their activities. Similarly, unless there is intervention to the contrary, air pollution created by vehicles driving in the streets is not a cost borne by the motorist.

The costs of air pollution on the urban society ranges from its detrimental health effects, to the need to clean buildings that are aesthetically damaged by such emissions.

2 *Noise pollution* can be generated from a variety of sources such as noisy neighbours that continually host rowdy late night parties, heavy industry and motorized traffic, for example, but it is perhaps the latter example which is the major creator of this negative externality in most urban areas. Again, it may only be with public intervention that some form of noise control could be placed on offenders. The motorist may not be too concerned about the noise from the engine of their car as he or she is stuck in traffic outside an office building, as it is the occupants of the building who are suffering the loss of welfare as they listen to *all* the traffic outside their place of work. Likewise, in a purely profit orientated, market world, the industrialist may not be too concerned by the noise levels of the productive process as long as that particular method of production maximized the firms profits. Such noise created by industry is not only detrimental and harmful to the people who work in the factory, but it could also be a nuisance to people who live nearby especially if the factory operated on a twenty-four hour basis. Even if the noise from the productive process itself was not so great, traffic noise created by the change over of 'shifts', and that of delivery vehicles arriving and leaving the factory gates could be a sufficient source of concern.

3 *Water pollution*. Again manufacturing industry is a good example of how water pollution can be perceived as a negative externality. Factories that discharge untreated, or partially treated, waste into a country's waterways can poison, or at least reduce the quality, of the river and thus damage the welfare of people who may derive a benefit from using the waterways. Such affected parties may include anglers and swimmers, but there may be other industries wishing to use the water from the river, but now have to incur the additional costs of 'cleaning up' the water so that it is of an adequate standard to be used for their productive process. Similarly, agriculture can lead to water pollution as surface runoff water drains large quantities of fertilizers into the river which can lead to the rapid growth of harmful algae killing off some forms of aquatic life.

4 *Traffic congestion*. Private road users obviously have to pay for the running costs of their vehicles such as fuel, servicing, insurance and depreciation. However, in a pure market economy, other costs of vehicle use that are imposed upon society would not have to be met by the road user. For example, the more people use their vehicles the more the urban environment will suffer from the associated problems of air pollution (see above). Moreover, congested streets also impose costs on the urban dweller in a number of other ways such as increasing the difficulties for pedestrians who wish to cross roads; increasing the number of vehicular accidents, especially those involving pedestrians; and creating vibra-

tional damage to buildings whose structures and foundations have to continually absorb the onslaught of heavy traffic. This latter point is especially important in the case of older, historic buildings whose structures may be weakened by the passing of time, but more importantly both their designers and builders could not possibly have envisaged the modern transport structure of the day and the damage that it can cause from associated factors such as acid rain and vibrational damage.

These examples are simply a brief examination of some of the main negative externalities common to the urban environment, and as per the normal range of examples given in this text, they by no means represent an exhaustive list. A useful exercise now would be for you to attempt to think of any negative externalities that may detract from your welfare in your everyday living in the urban context. Recognizing the fact that negative externalities exist, and can be considered as a problem for those living in our urban areas, one can now examine them in more detail so that policy can be formulated hopefully reducing or even eradicating such problems. In order to do this, the subject of economics gives us a simple model that should help us to quantify the extent of this problem. Although this model is extremely versatile and could easily be used to examine and illustrate all of the above examples, this part of the book will only deal with *two* examples due to the confines of space, but also to encourage you to attempt to utilize the model when examining other case studies of your own choice. The two examples that have been selected, one examining the case of urban traffic congestion, the other looking into the case of pollution in the urban environment, have been chosen to show the versatility of this model in its ability to tackle and analyse two seemingly diverse forms of negative externality. It is important to note that although the model attempts to quantify the extent of such externalities, this process is very difficult and unlikely to be fully accurate. For this reason the model's use is impaired, but it should not be dismissed as at least it highlights the fact that externalities exist and are thus worthy of further consideration and action.

Figure 15.1 can be used as the basis for both our chosen case studies. In dealing with the case of traffic congestion first, the diagram can be seen to examine the case of a busy urban road. The model depicted in the diagram looks at traffic flow per hour and the costs to both the private user and to society as a whole created by every *additional* vehicle that uses the road. The demand for the road by road users is shown by the demand curve *D*. This demand curve illustrates the benefits derived by the road users from its use. Such benefits will depend on the actual costs incurred by road users as determined by the various prices that the individual has to pay for all the items involved in road use. Therefore, the demand curve should reflect the 'marginal private benefit' (*mpb*) derived from such road use. Note that the demand curve exhibits the normal inverse relationship between price and 'quantity demanded'. In other words the higher the effective price of using the road, the less people will be encouraged to use it. Although if there are few alternatives to this route one would expect the demand curve

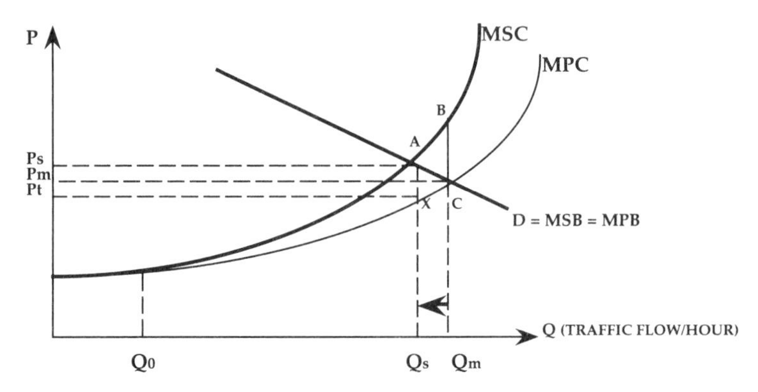

Figure 15.1. *The case of a negative externality*

to be highly inelastic. As society derives no additional benefits from such road usage the 'marginal social benefit' (*msb*) curve will be no different to the marginal private benefit curve. In other words: $msb = mpb = D$. The actual costs to the private road user are shown by the 'marginal private cost' curve (*mpc*). These costs not only include the costs of running the motorist's vehicle, but also reflect the cost in terms of time taken to travel along the road. That is, they also take 'on board' the notion of an **opportunity cost** to the individual as they have to spend time in their car in traffic rather than being elsewhere either at work or at leisure. At low levels of traffic flow per hour, for example Q_0, such private costs are relatively low as the journey is both quick and easy. However, as more and more vehicles use the road traffic builds up and congestion can occur. Once we have reached this stage private costs are likely to increase for a number of reasons. More time is lost at work or at leisure as the journey time is increased due to slow moving traffic, fuel consumption is increased as vehicles are rendered more inefficient by the continual stopping, or slowing down, on a busy road, and the physical wear and tear on a vehicle, notably the clutch, is increased under such driving conditions. Thus, if left to the market, road users would use this particular road until the cost of using it outweighed their demand for it. Such a point would be reached at *Qm* traffic flow per hour. Beyond this point traffic congestion would be deemed so severe that road users could seek alternative routes, use the road at less congested times of day, or reduce non-essential, marginal, trips. However, although the market system has arrived at a solution, many would argue that because of the existence of externalities associated with such high levels of traffic, such a solution would be an incorrect one. The argument is that the overall costs to society of the use of the road are different to those of the motorist as they also include externalities which are not captured by the market mechanism. Therefore, the costs to society of each extra vehicle using the road are *additional* to marginal private costs and are represented by the 'marginal social cost' (*msc*) curve. Because these marginal social costs are additional to marginal private costs the marginal social cost curve is higher than the marginal private cost curve. At low levels of traffic flow such as Q_0 the costs imposed upon society are relatively small as traffic is free flowing and relatively slight. However, as congestion builds up the additional costs of air pollution, increases in road traffic accidents, and 'nuisance value' of not being able to cross the road easily, are all examples of increasing costs upon society. The model illustrated in Fig. 15.1 would suggest that social costs increasingly diverge from private costs and would rise to a traffic flow such as *Qs* beyond which the number of vehicles using the road would be unacceptably high if one also took into consideration the costs imposed on the urban public at large rather than just the private road user. In other words, the differential between marginal social costs and marginal private costs becomes unacceptably high. Therefore, if the public sector's aim is to increase the welfare of society rather than the welfare of the individual, policies need to be implemented that would increase the cost of using this road from *Pm* to *Ps*. Such a policy of increasing the price of road usage should reduce the demand for that road as exhibited by the demand curve *D*. If this rise in costs was achieved the use of the road would be at the 'social optimum' of *Qs* rather than the 'private optimum' of *Qm* and therefore a **welfare gain** shown by the triangle *ABC* will be accomplished. This welfare gain is simply the saving in social costs that would have occurred if traffic flow were to go beyond *Qs*. A variety of policies could be attempted in order to increase the cost of road usage so that marginal private costs are increased so that they are more in line with marginal social costs. All of these policies are worthy of extensive debate but this is beyond the remit of this introduction to negative externalities. However, examples of potential policy are:

1 Charging for the use of the road at strategically positioned 'toll gates'. In this instance a toll of *PtPs* would need to be charged in order to reduce vehicular flow to the desired level of *Qs*. Such a charge would effectively close the gap between marginal private costs and marginal social costs.

2 A system of 'direct road pricing' (DRP) could be introduced in order to increase the costs of using the road. This would be achieved by either the use of 'off car' or 'in car' meters that would bill the driver of the vehicle for the use of the road.

3 Fuel price costs could be increased in order to indirectly increase the cost of using the road.

4 'Engineering solutions' can be attempted in order to encourage the demand for this congested road to decrease, preferably so that the new demand curve passed through point *X* because it is at this point that the desired traffic flow of *Qs* would be achieved. Such an engineering solution could be the construction of a ring road around a town that would remove the need for many vehicles to pass through the town and use the road in question.

Exactly the same model illustrated in this diagram could also be used to analyse the case of the negative externality of pollution. Imagine that there is a factory located within, or nearby to, the urban area that is producing goods, and that the manufacturing process it uses releases a degree of pollution into the atmosphere. The demand for the factories output is illustrated by the demand curve *D*. Again one would expect the normal inverse relationship between price and quantity demanded. That is, if the firm tried to charge too much for its product consumers would attempt to use less of it, or seek alternatives, and thus quantity demanded would fall when the price is set too high. The marginal private cost curve in this instance represents the costs of manufacturing the product. Such costs are likely to rise as output increases due to the 'law of diminishing returns' or 'law of variable proportions' (see Chapter 5). Thus, the market will tend to produce up to the point where the marginal private cost curve cuts the demand curve, and therefore *Qm* output is produced. However, because of the resultant creation of air pollution society may deem that such a level of output is too high – as it could be damaging to the health of local residents for example. In order to ascertain the socially optimal level of output, and therefore the 'optimal level of pollution', from this factory one would need to equate the demand curve with the marginal social benefit curve as this latter curve also includes the costs imposed upon the community by the negative externality of air pollution. Such a social optimum would thus be given by the lower output of *Qs* rather than *Qm*. Note that some pollution is acceptable to society as the marginal social benefit curve has already begun to diverge from the marginal private benefit curve by the time we reach *Qs*. These low levels of pollution are presumably acceptable as they pose no great danger to health, nor do they create any other significant problems.

(b) The case of a 'positive externality'

Positive externalities arise when the consumption of a good or service generates benefits to those other than the actual consumers themselves. For example imagine that an individual, or group of individuals, improved the quality of their own housing. Such improvements could be of the form of ensuring a high level of repair and maintenance of the buildings themselves, or the landscaping of the grounds that surround them. Although this may be done for their own private benefit such as improving their living environment, and increasing the value of their homes, it may also have positive spill-over effects that benefit society as a whole. In other words, expenditure on items for the purpose of improving the quality of housing in an area could enhance the visual and environmental aspects of the area for those who may visit, work, or pass through it, as well as improving the housing standards for the residents themselves. Just as with negative externalities, we can now examine the case of positive externalities by formulating a simple economic model that should enable us to go some way in quantifying the degree of the issue in question. Such a model is depicted in Fig. 15.2. In this figure it can be seen that in the case of a positive externality the marginal social cost curve is the same as the marginal private cost curve. Therefore, $msc = mpc$. The reason for this is that the production of

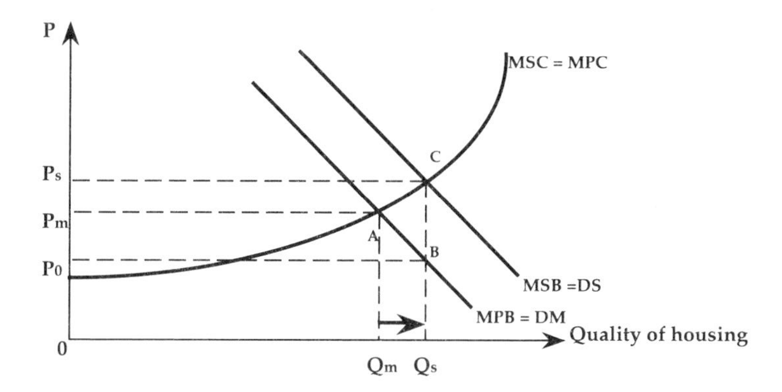

Figure 15.2. *The case of a positive externality*

such benefits imposes no additional cost to society itself and thus no adjustment needs to be made to the curve. The curve does rise however as improved housing standards can only be achieved at a rising cost. The demand for such home improvements by the home owners is represented by the demand curve *Dm* as this reflects the marginal private benefit of such improvements to them. Therefore, the individual owner occupier will invest in housing improvements until the marginal cost to him or her is equal to the marginal private benefit that he or she receives. Therefore, in this example home owners will tend to consume a house quality of *Qm* at a price of *Pm*. However, as argued above, the benefits to society of such home improvements are likely to be greater than the benefits simply accruing to the individual. This can be demonstrated by showing that society's demand for this work is higher than the demand of individuals. Here it can be seen that society's demand, as represented by *Ds* (*msb*), is significantly higher than the demand curve of individuals (*Dm*). Therefore, the problem is that if left to the market housing quality would only be as high as *Qm* although society would gain optimum satisfaction from a housing quality of *Qs*. Thus, it is argued that policy should be formulated to encourage individual home owners to improve their housing beyond the point where the marginal private benefit curve cuts their marginal private cost curve. This could perhaps be achieved by the granting of home improvement subsidies equal to P_0Ps so as to cover the extra costs required to bring housing up to the socially determined level. Effectively this would have the impact of raising the demand curve by the desired level. In this case society would experience a welfare gain equal to the area shown by the triangle *ABC* as these are improvements that would not have occurred without public intervention.

Urban policy and externalities

It has been recognized that if the market were left to itself to shape our built areas it could create an abundance of detrimental spill-over effects (negative externalities) that would reduce the overall welfare of people living in the urban environment. Therefore, one could argue that *carefully* designed and administered policy could be aimed at reducing or eliminating such negative externalities so as to improve upon the outcome of the market. Just as importantly one must also accept that the market, if left to itself, may not encourage a sufficient level of positive externalities to be created. Therefore, it is again felt that government policy could be aimed at promoting the maximum number of such positive spill-over effects so as to improve the environment of the urban dweller. Government intervention in this field can be at both the national and local level. So as to appreciate the type of policies that could be imposed a brief analysis of some of the main policy initiatives is examined immediately below.

(a) Taxation and subsidization

Governments could attempt to tax the producers of negative externalities. For example if there was a factory creating noise and air pollution due to the type of productive process that it used, the firm could be taxed to encourage it to switch to an alternative manufacturing process that created less negative spill-over effects. If the firm were to use a new, approved technique it would not attract the tax. Alternatively, the government could attempt to promote positive externalities by encouraging activities that could create positive spill-over effects. Such activities could be encouraged by making them more affordable with the assistance of publicly funded **subsidies**. For example if there was a part of the housing stock that fell below 'socially acceptable' norms, the government could offer subsidies such as home improvement grants to encourage the owners to improve their accommodation, in the hope that not only would such action improve the living standards of those who lived in the area, but would also increase the welfare of others in the community via the creation of positive externalities. These approaches are sometimes referred to as attempts to 'internalize externalities' as one is trying to incorporate externalities into the overall costs and benefits of society. For example: the internalization of negative externalities would try to bring marginal private costs in line with marginal social costs. However, there are difficulties with this approach to the externality issue. Firstly one needs to assess the relative desirability and undesirability of such externalities in order to decide which ones to take action on. Specifically people in society are not a homogeneous group as some will have different value judgements to others. Secondly, it is often difficult to place a monetary valuation upon externalities, and therefore assessing the correct level of tax or subsidy will be a resulting difficulty. Thirdly, one needs to ensure that the system of taxes and subsidies is both flexible and responsive to changing circumstances, and that it should also be relatively economical to run.

(b) The setting of standards

The government could aim to impose standards in the hope of improving society's welfare over time. For example if a polluting factory failed to meet current standards the authorities would be given the legal right to impose fines on such offenders. In a similar way, one could force firms to undertake activities that promoted positive externalities such as landscaping the surrounding area of their premises, and ensuring the upkeep of their buildings. If firms failed to reach such standards they could again be fined. The setting of standards for the construction of new buildings (building regulations) can also be used as a means of ensuring that negative aspects of poorly designed or badly built buildings are reduced, and that the positive aspects, such as easy access and good natural lighting, are maximized.

(c) Physical controls

Physical controls could be imposed by government to ensure that acceptable criteria are being met. The authorities could insist that anti-pollution devices, such as filters, could be fitted to factory's chimneys that were creating dangerously high levels of toxic emissions.

(d) Land use planning

The planning process could be used in order to remove the creators of negative externalities from certain areas of the built environment, and promote land use patterns that stimulated the maximum range of positive externalities in other areas. If the other measures listed above failed to eliminate the problem of pollution caused by industry in a certain area planners could either:

1 Remove the offenders, or generators, of the negative externality (industry) perhaps via the creation of an 'industrial park' in a certain area of the town or city that was sufficiently far away from competing land uses such as residential areas.

2 Remove the sufferers of the negative externality by relocating households to a cleaner more pleasant area of the built environment that was specifically zoned for residential and complementary land uses such as retailing and leisure.

Such relocation of land uses could be both a costly and time consuming process in an existing urban environment, yet would be relatively easy to achieve in the formation of a new town. Essentially planners would need to reduce the amount of conflicting land uses, and encourage complementary land uses wherever possible. Such a policy of 'zoning' can be seen in Fig. 15.3. This figure shows that in this hypothetical town residential areas have been placed away from the industrial sector. Importantly, such factors as the direction of the 'prevailing wind' have also been taken into consideration so as to ensure that, as much as possible, the pollution from the industrial area does not blow back on to the residential area and thus nullify the effects of the zoning policy.

(e) 'Bargaining'

In situations where the externality, the creators of the externality, and the parties affected by the externality, can be easily defined, a solution may be

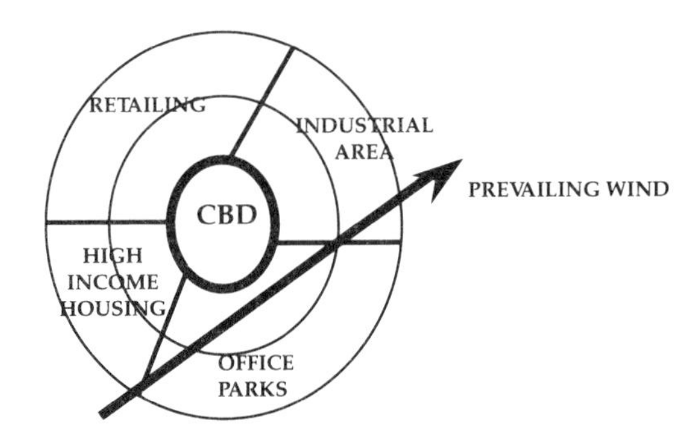

Figure 15.3. *Competitive and complementary land uses*

made possible by the concerned parties discussing the issue together. For example if one group is suffering from the effects of pollution from a particular factory, they may be able to negotiate some form of compensation from the factory owner. If their cause is a just one support can be given to the case from the legal system as the law is essentially designed as a form of protection for society. Moreover, consumers can give their cause considerable weight by organizing themselves – perhaps boycotting the firm's products if it does not agree to the demands of the people. Such consumer action has met with varying degrees of success in the past and largely depends upon the product in question. For example a heavy armaments factory will not be affected by a consumer boycott as it is the military rather than the general public who are the clients of such output.

In conclusion on the subject of urban policy and externalities it can be seen that a variety of policy measures can be, and have been, attempted in order to tackle the externality issue. Specifically the public sector is attempting to improve upon the solution given by the market. However, as suggested already, the public sector must ensure that their policies are sufficiently flexible and well administered in order to achieve this goal. Failure to do this could at best result in the policy failing to achieve its desired objectives or, at worst, the policy could inadvertently make the situation worse. There are also those who believe that the market will eventually take externalities into account and therefore such requirements reduce the case for intervention in the market. It is argued that in an area that suffers from a high degree of negative externalities house prices will reflect this by being lower than would otherwise be the case. Therefore, people are compensated for these detrimental spill-over effects by being able to purchase property at a low price. Conversely, areas that benefit from a high degree of positive externalities are often characterized by high house prices. Therefore, the market mechanism can be seen as a method of highlighting the externality issue through price. Moreover, there is always the suggestion that if an individual chooses to live in an urban area they must accept the existence of a certain level of negative externalities.

A conclusion on externalities

It is important to remember that externalities can be both positive and negative in nature, and that each category is equally important. It is also useful to appreciate that externalities can be identified and categorized into four main types depending upon the source of the externality and the party which it affects. These categories are listed and briefly described immediately below.

(a) Producer/consumer externalities

These are perhaps the most frequently discussed externality primarily because they are normally very noticeable. An example of such an externality is our now well-used case of the factory polluting the atmosphere. Here we have a producer (the factory) creating a negative externality that adversely effects the welfare of the inhabitants of the area around the factory because of air-borne waste. Likewise if the factory was causing the pollution of waterways it would reduce the pleasure derived by people who use such waterways for recreational purposes such as swimming or boating. Similarly one may argue that an architect in designing and constructing a building that is not in keeping with its surroundings, may impose the negative externality of an aesthetically unpleasing 'eye sore' which is not only unpleasant in itself, but could also detract from the whole surrounding area. Producer/consumer externalities need not only be negative in nature, however. An example of a positive externality being created by a producer to the benefit of the consumer is perhaps seen in the world of agriculture. Here the farming of the land creates a well-managed countryside that is pleasant to see and to walk in.

(b) Producer/producer externalities

Producer/producer externalities occur when the production of one producer adversely affects the production of another. The pollution created by heavy industry could reduce the yields from surrounding agricultural land, or pollute nearby waterways. Again we can perhaps imagine a positive externality under this heading. If there were two office blocks adjacent to one another and one building was to undergo extensive external refurbishment it is likely to improve the look of the area in a way that could also attract clients to the unimproved building.

(c) Consumer/producer externalities

Consumer/producer externalities occur when the actions of the general public has an impact upon the welfare of producers. Ramblers and dog walkers who visit the countryside may inadvertently wander off the designated footpaths and cause damage to agricultural fields. Such activity will obviously damage the farmers' return from such fields. Similarly, the industrialist may have the problem of ensuring sufficient access for lorries into the factory if nearby roads are clogged with parked vehicles of workers in other factories or offices on nearby access routes.

(d) Consumer/consumer externalities

These occur when the behaviour of one individual, or group of individuals, affects the welfare of another individual, or group of individuals. An example of consumers creating a negative externality would be the neighbour continually hosting noisy late night parties. Alternatively, an example of a positive externality under this classification would be the home owner whose garden was carefully landscaped and/or house kept in an immaculate state of repair. Such actions are likely to improve the welfare of all who see the house rather than just that of those who live in it.

(2) The case of 'public goods' in the built environment

Public goods are goods and services that are certainly valued by individuals, but they are unlikely to be provided by the market mechanism, or at least in sufficient quantities. Common examples of public goods in the urban context are public open space, such as urban parks, and street lighting. The reason that the market does not tend to cater for such items is that they arguably exhibit both the characteristics of 'non-rivalry' and 'non-excludability'. **Non-rivalry** implies that one person's consumption of a good does not impair the potential utility that another individual can derive from that good. For example if one person walks past a street light the amount of light available to others is not reduced in any way. Likewise one could argue that your enjoyment of the local park is not affected by other people using the facility. More significantly though, public goods also exhibit non-excludability. **Non-excludability** implies that once one provides the good one cannot exclude anyone from using it, and if this is the case it makes it difficult, if not impossible, to charge a price for it. For example, once street lights are turned on there is no way that you could limit their use solely to those people who had paid a contribution to the service. In other words, the problem of the **free rider** can occur, where a free rider is an individual who derives a benefit from a good or service which they have not paid for, while others have. It is chiefly for this reason that the only way that such services can be financed is via a legally enforceable system of taxation rather than direct charges, and it is only the public sector which has the power to make such charges. In order to clarify this issue the case of public goods can be directly contrasted with that of private goods. **Private goods** exhibit both rivalry and excludability inasmuch as you can exclude others from using your private property and your enjoyment of your private property largely excludes others from using it. Moreover, as you can determine the use of a private good, you could charge for its use if it were to be used by others. For example, your home or your own private car are illustrations of these points. Therefore, it would appear that in order to ensure that our urban areas contain a sufficient quantity of public goods direct government intervention in the market is required to raise the finance for their provision. Thus, it would seem that town planners must decide upon how much urban road space to provide, how many areas should be designated as open public space, how much street lighting should be provided, and so on. In order to do this planners must accurately determine the demand for such goods so that the correct allocation of them is provided in terms of attempting to maximize society's welfare. However, just as there are supporters for the view that the market can address the externality problem, there are those who argue that the assumption that certain goods should be classified as public goods is not necessarily a valid one. The proponents of such a view stress that for a good to be a pure public good it would need to fully exhibit the dual characteristics of non-excludability and non-rivalry. However, finding goods with both of these characteristics is relatively difficult, and if this is the case the argument for public goods begins to weaken. For example urban roads are normally seen as being the responsibility of local government as they are assumed to be public goods. Although if one examines the issue more carefully, one could arrive at the conclusion that urban roads do not conform to either of the characteristics necessary for the classification of a public good. With respect to non-rivalry it is not true to say in this case that one person's use of the road does not effect the utility of others using the road, as it is quite clearly the case that, as congestion begins to build up, using the road becomes slower and more dangerous to name just two negative aspects. More importantly, to be a pure public good, it must adhere to the characteristic of non-excludability. However, it is possible to exclude certain people from using the road by either charging a toll or only admitting vehicles with a special licence, for example. Once one can do this one is able to charge a price for its use, and therefore one can see the opportunity for the private sector to get involved in the provision of what has traditionally been assumed to be in the public domain.

(3) Monoply situations in the built environment

Monopolies are most commonly associated with the production of goods and services (see Chapter 7, however the issue of a monopoly can occur in the case of urban land and development. For example, imagine an area of land, such as that shown in Fig. 15.4, that is currently made up of four individual plots each with an existing building, and owned by four separate owners. It may be the case that this block of land is the ideal site for the development of a new, larger building that would replace the existing structures. If the existing buildings were in a poor state of repair, were dilapidated or were functionally obsolete, and had no historical interest, it may well be in the interests of the urban society that the new development were allowed to go ahead. In order for the new building to be constructed the developers would need to acquire the block of land in total and would thus need to negotiate selling prices of the existing buildings and sites with the current owners. It may be relatively easy at first to buy up the plots as each existing owner is in competition with the other land owners to sell

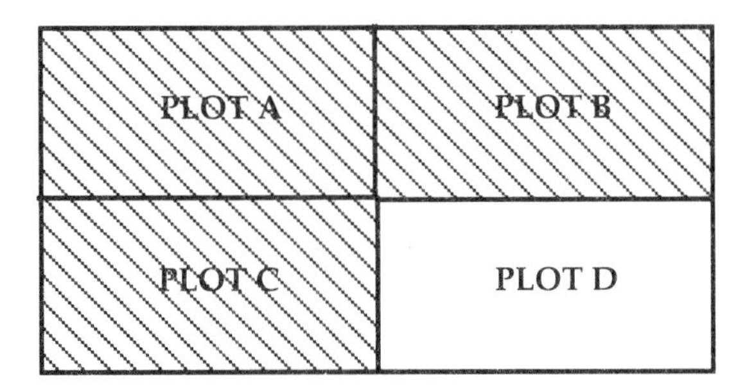

Figure 15.4. *The monopoly landowner and redevelopment*

their property. So, for example, landowner *A* may not wish to ask for too high a price just in case he or she loses the sale to one of the other owners *B*, *C* or *D*. In fact such behaviour is likely as long as there is competition to sell amongst owners, and as long as the existing landowners are unaware of the developer's overall intentions of purchasing all four properties. However, if the developer has managed to purchase all the plots except plot *D*, for example, and the owner of that plot knows that the developer needs that plot in order to make the development viable, he or she could hold out for a *ransom price*. In other words, the owner of plot *D* could ask a price well in excess of the selling price of the other three properties as this remaining landowner is now in a strong monopoly position. In fact, according to the above logic, it is in every one of the property owners' interests to be the last seller in such circumstances as they try to hold out for the higher price, and there is therefore the danger that no one will sell at all. Thus, the market may not be able to cope with this problem and 'monopoly landowners' could be in a position to delay or prevent a beneficial project being undertaken, or one may gain significantly more from their property than those who sold previously in competition with one another. Therefore, it is argued that there is a need for public intervention, such as the power of a compulsory purchase order (CPO), which is designed to enable a local authority to forcibly purchase properties that are required for a particular worthwhile development. Thus, such a system could be designed to facilitate the assembly of units of land and property that are presently in fragmented ownership. Without such powers many feel that much urban investment and redevelopment would not take place at all. However, planners must ensure that such powers are not granted too hastily as a solution could be achieved via the refurbishment of the existing buildings rather than via their demolition. Perhaps one of the reasons that many cities have lost fine examples of historical architecture is that planners have too readily redeveloped areas without much consideration to the conservation of the built environment. Another argument against the granting to the state of such interventionist powers is that they intervene too directly with the freedom and wishes of the land or property owner. A property owner may not wish to sell in the current market if prices are depressed, and thus they may be waiting to sell in more prosperous times where a higher selling price can be achieved and therefore more profits can be made. However, the serving of a compulsory purchase order would force the owner to sell at the current, undesirable, market rate.

Also in relation to the theory of the firm, planning powers may also be used to try to ensure a socially desirable mix of buildings on a site. For example, the owner of a vacant plot of building land set aside for residential development may find that it is likely to be more profitable to restrict the development to a few highly priced 'executive style' houses rather than any other form of residential buildings. As seen in Chapter 5 such profit maximization would occur when the builder's marginal costs were to be equal to their marginal revenue. Therefore, by referring to Fig. 15.5, it can be seen that if the market were allowed to dominate, the development will be restricted to the construction of Q_A houses that would be sold for a price of P_A. However, the needs of society may not be met by the building of such houses as they will only cater for people on high incomes, or for companies who wish to purchase such houses for their senior management for example. The requirements of the local community may be better met if the development consisted of smaller houses or a 'mix' of types of houses that included smaller units such as starter homes. Therefore, the planners should recognize the undesirability of the market solution in such instances and insist on the development of Q_B houses rather than Q_A. The smaller houses would have a lower, and subsequently more affordable, selling price of P_B.

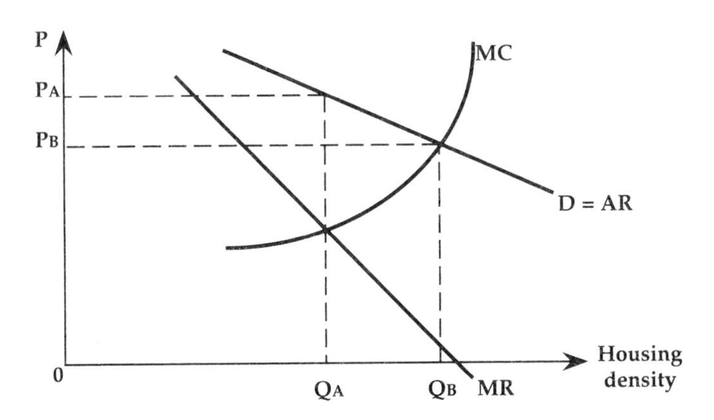

Figure 15.5. *The private sector development decision*

(4) Imperfect information in the urban market

If a market were to operate perfectly all those in the market, both buyers and sellers, would require perfect information concerning the ruling level of prices, the behaviour of competitors, and so on (see the theory of Perfect Competition in Chapter 7). However, in the absence of any public intervention in the built environment such perfect knowledge is extremely unlikely, and if this condition of perfect knowledge does not hold the results produced by the market mechanism are unlikely to be satisfactory. When examining perfect knowledge in the case of the *producer*, one can envisage two potentially damaging scenarios that could occur:

1 Developers could *underestimate* the demand for a particular type of building so that the urban area becomes inadequately serviced by certain land uses. If the developer did not appreciate that a town was going to experience an increase in wealth or an increase in migration, they may not provide sufficient retail areas. Therefore, at least in the short run, there will be increased pressure on existing facilities that may lead to overcrowding and congestion just to mention two detrimental effects of over usage.

2 Developers may *overestimate* the demand for a particular type of building. If a development firm notices that its rivals are constructing and successfully selling office buildings in a town, they may also wish to take advantage of such potential profits. However, if such behaviour is left unchecked one can easily envisage that the market will become flooded with a particular type of building. Such an oversupply will tend to drive down rental levels of the existing buildings as clients have greater choice, but it could also mean that many buildings remain unoccupied and are at risk of becoming derelict. Vacant buildings tend to attract a whole host of other social problems such as vandalism and other related crimes. Alternatively one can imagine the example of a developer developing a large retail centre, but is unaware that another similar facility is going to be built by another firm in the same vicinity in the near future. If both retail centres are built consumer demand may not be sufficient in the area to ensure the prosperity of two facilities. Just as with the case of the oversupply of office buildings, one can see here that many retail units could fail or remain unoccupied and that this would reduce the benefits derived from such centres. However, supporters of the market system could respond here by arguing that such competition will cause a 'price cutting war' amongst producers that will be to the benefit of consumers. This argument is often backed up by the view that restrictive planning creates unnecessary scarcity which in turn causes urban prices to remain high.

Advocates of the market mechanism may also argue that the market would eventually arrive at the correct provision as the weaker developments would fail in the long run leaving the urban area with the correct quantity of each type of land use. Buildings that were originally intended for office use, but remained unoccupied, could perhaps be converted in order to accommodate a change in use and be used for another purpose such as education. However, this argument perhaps fails to appreciate that the short-run adjustment period may be quite a long one, and that firms may not supply less profitable buildings that are still of value to urban society. Therefore, there is a case for public intervention in the form of planning, whereby planning would be used in conjunction with the operation of the market in order to achieve a socially acceptable range of developments. It is likely that planning departments have the best knowledge of the future requirements of the urban area as they are also involved in the planning of residential areas and access routes, which both have a bearing on the location and success of other land uses such as commercial and retail centres. By definition planners need to forecast the future growth and changes in the urban area, and such studies and ongoing monitoring is invaluable in relation to ascertaining the correct number and location of future developments. Again the supporters of the free market system may react to this view by stating that this argument for centralized planning assumes a lack of sophistication on behalf of private firms. If private development companies wish to ensure their success and maximize their profitability it is presumably likely that they too would attempt to forecast the future requirements of an urban area before committing themselves to any development. In fact many large development firms do indeed have their own forecasting personnel or departments. However, this standpoint still does not address the provision of less profitable land uses. Moreover, it is still unlikely that the private sector would be able to obtain as much information as the public sector with respect to the overall view, and therefore it would seem best that the two were to work in conjunction with one another.

Imperfect information in the market can also be a problem for *consumers*. A firm of retailers wishing to open an outlet in a town may be unaware of the plans of their competitors as other firms may also wish to set up in the vicinity. Therefore, in a pure market situation this firm could be expecting a highly profitable location, but in the light of subsequent competition the site could turn out to be highly marginal or even unprofitable. Such uncertainty could put investors off a town which could result in a loss of potential jobs and desirable land uses. With a planning system in force, however, the firm could look at the overall future structure of the urban area and be able to make more certain decisions. Moreover, the planners would know how many other applications they had for similar projects. Another example of imperfect information and the consumer can be seen in the case of the house buyer. With reference to town plans a potential house purchaser can quickly see the other types of land use that are in the immediate area of a house. Such knowledge could prevent house buyers buying a house that will soon be adjacent to potentially undesirable land uses such as heavy industry. In a pure market situation such information would not exist in an easily accessible, centralized form. Instead the location of development would depend chiefly upon the profitability decisions of individual firms, and such decisions and locations could change rapidly in response to continual changes in the market giving rise to an air of uncertainty.

(5) Social views and the urban environment

Many argue that the preferences of the individual expressed through the market mechanism are not the same as the views of society as a whole. The simple explanation of this view is that one will try to maximize one's own welfare without too much regard for others in society. It is because of this view that planning can be seen as adopting a 'paternalistic policy' which guides society to make decisions that will maximize the welfare of the whole urban community rather than that of the individual. There are those who believe that if everyone maximizes their individual welfare this should compound to the maximization of society's welfare. However, this apparently logical view fails to recognize that the market only caters for those who can pay (the whole notion of effective demand), and therefore there are bound to be sectors of the community, for example the poor and homeless, who will not be included in such a system, and subsequently such people could not benefit from its operation. Moreover, people in society may not fully appreciate the value that they derive from certain land uses such as open public space, and therefore if left to the individual, urban areas could suffer a lack of certain important amenities, say public parks. Similarly, planning permission for a residential development could be refused if the planners felt that individuals had failed to recognize the full significance of the nuisance value created by nearby industry, or a major road. Furthermore, society may not fully take into account the needs of future generations. The preservation of historical buildings leaves an important record of architecture and the development of communities over time. A purely market orientated urban environment may demolish such buildings if they are no longer seen as being profitable. Finally, it is argued that society will not cater for a sufficient level of provision of 'merit goods'. **Merit goods** are those goods which it is felt that society should consume to an adequate level in order to ensure an acceptable level of welfare. Examples of such merit goods include schools, libraries, and hospitals.

A conclusion on market failure and the rationale for public intervention in the urban market

It can be seen from the above analysis that there are arguments both for and against the market as being the sole 'driving force' behind the creation of the built environment. Generally it would appear that there is some merit in instigating a degree of intervention in the market in order to correct its inherent shortcomings. Without such intervention the welfare of society as a whole is unlikely to be maximized. However, as mentioned above, such intervention would have to be managed carefully and flexibly in order for it to achieve its desired results. Research has revealed that there are many fundamental and common criticisms of the planning process, and indeed other forms of public intervention in the built environment. Problems with the planning process include:

1. It is argued that planning is all too often used in a *negative* manner that prevents development. Although one of the roles of planning is of course to prevent the construction of buildings that are unsightly or not required, many feel that planners rarely take on their other role of positively promoting worthwhile development.
2. Many complain that the planning process is a highly bureaucratic one that leads to the slow processing of case by case applications. Such time lags can be highly costly to the developer as it lengthens the period of development and the potential need for borrowed funds. Therefore, developers costs can be increased and the receipt of their revenues delayed as the sale of the completed building is postponed because of delays in obtaining planning permission at the early stage. Such difficulties could well lead to the cancellation of investment in the urban environment, and any cost increases from delay mitigates against the development of marginal projects.
3. Another common argument is that some of the guidelines of planning are too broad and lack sufficient flexibility to accommodate individual cases with individual circumstances. Although it is felt undesirable to have a certain type of building in a particular area of town there may be circumstances where it could be seen as acceptable. A 'ban' on industry near a residential area perhaps needs to ensure that it differentiates between the cases of 'light industry' as opposed to 'heavy industry'.
4. Some studies of town planning suggest that planners sometimes fail to appreciate certain effects created by their actions. They may have overlooked the fact that their decision has had a 'distributional impact' upon an area of the town. It may be the case that giving permission for the development of a new retail centre encourages higher income groups to live in that area, and this causes house prices to rise beyond the reach of poorer members of the community who wish to live in that area because their family and friends have been long-run residents there.

Therefore, just as there are criticisms of the market, there are equally strong arguments against various forms of state intervention. However, hopefully it can be seen that the problems of the planning process above could be rectified by making such intervention more efficient and 'streamlined'. Finally it must be re-emphasized that one should examine the relative merits of all systems and hopefully take the best from each. If we adopt this view we are likely to see the public and private sectors working 'hand in hand' in achieving the maximum possible welfare for urban society.

CONCLUSION

This part was designed to help show how the built environment is created with inputs from both the public and private sectors. Importantly in times of declining direct public involvement in construction and its related industries in many countries the part highlights the merits of private sector investment, but also examines the potential failures of the market system. Again, with the information gathered here one should be able to explain existing features of the built environment such as urban land use, as well as changes in that environment as a range of variables influence it.

SUGGESTED QUESTIONS AND TASKS

The following are a number of questions that could be used either as the basis for discussion in tutorial groups, or as examination or coursework essays. The majority of the questions are suitable for the individual to approach, although some may be best suited to a group exercise. The relevant chapter is given at the end of each question. Group exercises are recommended to promote the student's ability to work in a team, and to enhance a student's skills of organization and management when producing a joint report. Verbal presentations are encouraged to foster confidence in dealing with people and presenting material to large audiences.

1 Analyse why house pricees can vary from one region to another within a country. (Chapters 1, 2 and 3a)

2 Statistics collected over a period of time indicated that house prices and the number of houses purchased and sold had both increased over the years.
Discuss whether or not this information implies an upwards sloping demand curve for houses, and analyse other potential solutions to the problem. (Chapters 1, 2 and 3a.)

3 Discuss how a knowledge of demand, supply, and elasticity coculd assist a building firm in estimating the likely final prices for their products. (Chapters 1, 2 and 3.)

4 'Due to its inelasticity of supply, land prices are increasing rapidly, and this can only lead to higher house prices.'
Discuss the validity of this statement. (Chapters 1, 2 and 3.)

5 Using 'market analysis' discuss why rents for both commercial property and industrial property change over time. (Chapters 1 and 2.)

6a Assess the arguments that may be put forward for introducing a 'price ceiling' on rented accommodation in your home town.

6b Discuss what secondary measures, if any, that you could impose to reduce any possible 'ill effects' of such legislation. (Chapters 1 and 4.)

7 Examine the case for government controls on land use. (Chapters 1 and 15.)

8 After undertaking extensive market research, and obtaining planning permission to build a new office block, the office developer finds that as

building height increases the market value of the floor space created on a particular site increases less than proportionately to capital development costs. The details of the developers survey are shown in the table below:

A	1	2	3	4	5	6	7	8	9	10
B	300	120	120	120	120	120	120	120	130	140
C	600	900	1120	1310	1485	1632	1757	1867	1965	2052

where A = number of floors,
B = the capital cost per floor,
C = The total value of the total floor space created.
All figures for *B* and *C* are quoted in (£000s).

a If profit maximization were the goal of the developer, show clearly via both tabulation and an explanatory graph what height the developer should build the building. Give full justification to your answer.
b Discuss why the developer may not be able to achieve the profit maximizing solution, by highlighting the constraints that may exist on the development.
c Suggest reasons why the decision made in (8a) may vary during the actual construction of the building. (Chapter 5.)

9 You are consulted by a house builder who estimates that for the construction of 'starter homes' costs will be as follows:

A	1	2	3	4	5
B	24	44	74	112	170
C	20				

where A = houses to be built;
B = total variable costs;
C = total fixed costs.
All figures for both variable costs and fixed costs are quoted in (£000s).

a If each house can be sold for £55,000, carefully advise the building firm on their best course of action.
b If there were to be a period of sustained high interest rates, for example, leading to depressed demand, what would your advice be to the builder if the maximum price that the houses could be sold for dropped to £30,000. (Chapter 5.)

10 Examine the assertion that a firm of builders should only build up to the point where 'their marginal cost curve cuts their marginal revenue curve from below'. (Chapters 5 and 6.)

11 'The larger a firm of surveyors becomes, the more efficient it will be.' (Chapter 6.)
Discuss this statement.

12 Assess the implications to construction firms of declining demand for their output if they are operating in a perfectly competitive environment. (Chapters 1 and 7.)

13 Assess whether the existence of monopolies on the materials supply side of the construction industry will be an overall disbenefit or benefit to construction firms. (Chapter 7.)

14 Imagine that:

a A property developer had acquired a two and a half acre plot of land for £250,000 per acre.

b The site is in a prosperous suburb near to good road and rail communications.

c The money for the land was raised via borrowed funds at a rate of interest of 10 per cent.

d The property development company wishes to maximize its profits.

e Planning consent has been given for the site to be developed for residential use, with no additional guidelines or constraints, except that no building should be higher than three floors.

Now imagine that your group has been hired by the development firm to advise them on how to proceed with the development. Using the information above devise a development that will achieve the desired goal. Your joint report should be about ten sides of A4 in length and contain sketch maps of your proposal as well as a full written description. Care should be taken to ensure that your scheme has a good foundation of logical economic theory based upon market analysis, and data obtained from similar developments and housing stock in your area. You must be able to justify your site layout, the type of houses selected for the site, and the potential market price for the completed development. (Chapters 1–8. This question is suitable as a group exercise.)

15 After formulating a simple macroeconomic model demonstrate the implications of the following on both the economy and the property market.:

a An increase in the level of public spending in the economy.

b An increase in the value of the domestic currency on international exchanges.

c A deterioration in this country's balance of trade.

d An increase in the level of income tax.

e A decrease in interest rates.

f An increase in the marginal propensity to consume, (*mpc*). (Chapter 9.)

16 Explain carefully what is meant by the term the 'Multiplier Process', and via the use of examples and specific reference to property, demonstrate how this process can be both 'positive' and 'negative' in nature. (Chapter 9.)

17 Describe what is meant by the term 'expansionary demand management', and assess how such a policy could affect both the building industry and the property market. (Chapters 9 and 11.)

18 Examine the implications for two of the following property markets of a contractionary monetary policy.

a Retail property;
b Residential property;
c Industrial property;
d Leisure property;
e The office market. (Chapters 9 and 11.)

19 Examine the potential failures of fiscal policy and discuss how such problems can be reduced or eliminated in order to increase the effectiveness of this form of demand management. (Chapter 11.)

20 'Relaxed monetary control can only lead to inflation in the property market.'
Discuss this statement. (Chapter 11.)

21 For the long-run stability of the macroeconomy, demand side policies should be replaced by supply side policies.
Discuss this view. (Chapter 11.)

22 Examine how government controls on the financial institutions could affect the price of residential property, and the volume of new house building. (Chapter 11.)

23 If the government fails to meet its economic objectives property markets will be adversely affected.
Discuss. (Chapter 10.)

24 Examine the key forces of urbanization, decentralization, and potential gentrification that have occurred, or are still occurring, in your local largest town or city. You are also required to highlight the impact of such changes upon the rate of redevelopment and refurbishment, and to suggest what social costs and benefits these may lead to. (Chapters 12 and 14.)

25 Due to the recent opening of a new motorway a few miles to the west of a nearby town, there has been a substantial build-up of traffic crossing an existing bridge during peak travelling times. This has caused such severe congestion that the city council, because of the lack of public funds, have allowed private contractors to look at ways of improving the river crossing. In exchange for improving the crossing the private developers would be allowed to charge a fee to each motorist making use of their facility so that an income can be generated and profits can be made.

To date two firms have submitted proposals for the scheme. The first firm favoured a bridge (Project A), whilst the second firm favoured a tunnel (Project B). Both projects are seen to be comparable as initial capital costs are forecast to be nearly the same.

If long-run profit maximization is the overriding motive of the firm, use the net present value technique of investment appraisal to show which project would yield the highest returns. Assume a value of $r = 10$ per cent, and an anticipated project life of ten years.

YR	0	1	2	3	4	5	6	7	8	9	10
A	1000	220	220	220	100	100	100	100	100	100	60
B	1100	190	190	190	180	180	180	180	180	160	160

where A = project A;
B = project B.
All figures are in (£000s).

a Given the above information, state which project is the most viable.
b Suggest reasons for the patterns of cash flows observed for both projects.
c Suggest reasons why the results of your NPV calculations are unlikely to be fully accurate.
d Diagrammatically estimate the Internal Rate of Return (IRR) for either Project A or B. (Chapter 13.)

26 Assess how the theory of the Economic Life of a Building can be used to determine the economic life of the main retailing areas of the central part of your home town given the following set of circumstances:
a The growth of 'peripheral' retailing centres on the outskirts of the town.
b A growth in the population of the town.
c An increase in the cost of repair and maintenance work.
d A reduction in the rate of interest.
e The imposition of a tight monetary policy in conjunction with deflationary fiscal policy by central government.
f The charging for car parking in all areas around the central shopping area.
g A rise in the price of energy. (Chapter 13.)

27 Assess how Cost Benefit Analysis (CBA) can be used by the public sector to justify the creation of a bypass around an existing village that is currently characterized by a major through road between two large urban centres. Imagine that the village has a large number of buildings of historical interest, and that small traders dominate the high street. The surrounding land around the village consists of both grazing land and woodland. (Chapter 14.)

28 Examine the assertion that the property market is both a 'barometer' and 'regulator' of the economy as a whole. (Chapters 9–11.)

29 A national firm of house builders is concerned that its profits are being eroded by low productivity. You are required to write a brief report of about ten sides of A4 in length, suggesting why productivity could be low on their sites. More importantly, you are invited to put forward well-argued suggestions and methods that could increase the level of site productivity and innovation. It is important that you visit local house building sites in your home town in order to gather supporting data and evidence. (Chapter 8. This question is suitable as a group exercise.)

30 Imagine that you have been invited by a housing association to present a brief report/paper on the financial viability of increasing the level of investment into energy saving devices in their housing stock. You are required to select an appropriate investment appraisal technique to demonstrate whether the installation of a range of energy saving devices would represent a significant financial saving for the housing association or not. Moreover, you should briefly justify why the investment appraisal method that you have chosen is the most suitable in this instance. Furthermore, the choice of energy saving devices is up to your own discretion. (Chapter 13. This question is suitable as a group exercise.)

Index

Agglomeration economies, 72–73, 191, 195, 197, 224
Aggregate:
  demand, 147, 149–150, 163, 164, 166, 175,177
  supply, 148, 149–150, 177
Aid-tying, 136
Ando, A. 179
Anti-monopoly legislation, 115
Arc elasticity, 30
Areas of outstanding natural beauty, 72, 202
Automatic stabilizers, 173
Average costs, 92, 94–95, 97–102
Average rate of return, 209, 211–212
Average revenue, 90

Backwardness, 130
Balance of trade, 145, 157, 160–161, 167
Balanced budget theorem, 170
Barriers to entry, 115, 118, 135–136
Bid rent curve, 191
Brain drain, 148
'Break even', 96
Bricklayers, 60–65
Budgets, 164, 172
Building:
  boom, 58–60, 64, 145, 162
  costs, 90–95
  labour, 60–65
  land, 52–57
  materials, 57–60
    fashion, 58
    price, 58,
    price controls of, 78–79
    quality, 57
  recession, 63–64
  regulations, 243
  societies, 50, 172
  standards, 57, 72
  supplies, 20–21
  wages, 62
Bulk purchase discounts, 98
Business rates, 185
  park, 72

Capital:
  equipment, 132
  expenditure, 165–167
  grants, 165, 185
  intensity, 132
  labour ratio, 131–132
Cartels, 104
Central business district, 192
Central planning, 1
Ceteris paribus, xiv, 7
Change in use, 207, 219, 221
Closed shop, 182
Commercial property, *see* Office buildings; Retail property; Industrial property
Complementarity, 38–39, 43–44
Complementary goods, 12, 14–15, 38
Complementary land use, 72, 224
Compound interest formula, 212
Comprehensive redevelopment, 204
Compulsory purchase orders, 54, 224, 248–249
Computer aided design (CAD), 100
Confidence, 159
Conflicting land use, 224
Conglomerate merger, 118
Conservation, 194,199, 203, 207
Construction:
  components, 57–60
  industry, 122–138
Consumer:
  demand, 143–146
  subsidies, 70
Consumption, 143–146
Contiguous urban areas, 189
Contractionary fiscal policy, 167–168
Contractionary monetary policy, 175, 176
Conversion, 202, 207, 225
Conurbations, 189
Cost benefit analysis, 227, 228–233
Costs, 90–102
  average costs, 92, 94–95
  fixed-costs, 91, 93–94, 227
  long-run, 97–102,
  long-run average total costs, 97–102
  marginal costs, 92–93, 94–95
  short-run, 90–95
  short-run average total costs, 97–102
  total costs, 92, 93–94, 227
  variable costs, 91–92, 93–94, 227

Credit, 144
  availability, 173, 174, 175, 176
  restrictions, 176, 178
Crowding-out, 171–172
Current expenditure, 165–167
Current income, 178
Cut-off line, 229
Cyclical growth, 162–163

Decentralization, 194, 196–205, 222
  commercial, 200
  housing, 198–200, 201, 203
  manufacturing, 196–197, 202–203
  retailing, 200, 202–203
Decision lags, 168
Deflationary effect, 167–168
Deflationary gap, 164, 173
Demand, 1–17
  complementary goods, 14–15
  curve, 7–8
  declining, 27
  derived demand, 55, 60, 146, 174.
  determinants, 3–4
  elastic, 33–35, 60–61
  excess, 24
  functions, 4
  income, 9–10
  inelastic, 35, 60–61
  management, 151, 163–179
    fiscal policy, 163–173
    monetary policy, 173–179
  price, 7–8
  price elasticity, 29–41
  substitutes, 12–14
  tastes, 15–16
  total demand, 16–17
Demography, 49, 145, 174, 199
Depreciation, 137
Depression, 162, 164
Dereliction, 72, 151, 158, 164, 194, 202
Deskilling, 133
Devaluation, 137
Development mix, 249
Differential subsidies, 76–77, 79
Diminishing returns, 93, 97
Direct road pricing, 239
Discontinuous urban areas, 189
Discounting techniques, 213–216
  internal rate of return, 215–216
  net present value, 213–215
Diseconomies of scale, 97, 100–102
  labour, 101
  management, 100–101
  markets, 102
  raw materials, 101
Disposable income, 9–10, 69, 166
Distribution of income, 145
Division of labour, 133
Divorce rate,199
Duopoly, 104
Dusenberry, J., 179

Economic life of buildings, 57, 217–225
Economic growth, 157–158, 162
Economic model, xii
Economic rent, 62–63
Economies of scale, 97, 98, 124
  in finance, 99
  in management, 98
  in raw materials, 98–99
  in research, 99
  in specialist equipment, 99–100
Effective demand *see* Demand
Efficiency, 25, 159
Elasticity, 29–44
  arc, 30
  broadness of category, 40
  complementarity, 38–39
  cross elasticity, 43–44
  demand, 29–31, 33–35, 39
  determinants of, 38–41
  housing, 46–48
  income, 39–41
  income elasticity, 41–43
  inelasticity, 35–37,
  perfect, 34
  point, 29–30
  price, 29–41
  relative, 33–37
  substitutability, 38–39
  supply, 31–32, 33–35, 41
  time period, 40
Energy savings, 67, 68, 69, 206
Entrepreneurs, 148
Environment, 202
Equilibrium, 23–26
  automatic, 25
  efficient, 25
  stable, 25
Equity, 25, 52, 158, 159
Error term, 4
Estate agents, 51
Excess:
  capacity, 150
  demand, 24, 144
  supply, 25, 158
Exchange:
  controls, 137
  rate, 147, 160
Execution lags, 168–169
Exogenous shocks, 158
Expansionary fiscal policy, 164–165

Expansionary monetary policy, 174–175
Expectations:
  in demand, 8–9, 11, 144
  in supply, 20–21
Exports, 147, 156, 161
Externalities, 234–246
  bargaining, 243–244
  consumer-consumer, 246
  consumer-producer, 245–246
  housing, 51–52, 227
  internalization, 242
  land use planning, 243
  negative, 235–240, 241–245, 246
  physical controls, 243
  positive, 235–240, 245, 246
  producer-consumer, 245,
  producer-producer, 245
  standards, 242–243
  tax subsidies, 242
  urban policy, 241–243

Factor prices, 112
Financial economies of scale, 99
Financial penalty clauses, 130
Fine-tuning, 164, 172
Fiscal policy, 49, 163–173
  contractionary, 167–168
  crowding-out, 170
  expansionary, 164–165
  fiscal drag, 170
  limitations, 168–172
  opposing multipliers, 170
  pump-priming, 176
  time-lags, 168–169
  uncertainty, 169–170
Fixed costs, 91, 93–94, 227
Fixed price contracts, 129–130
Flow, 47, 52
Foreign currencies, 137, 147, 160
Free-rider, 246
Frictional unemployment, 182–183
Friedman, M., 179
Functional obsolescence, 54, 196, 203, 210, 217

Gentrification, 199–200
Globalization, 122–123, 138
Government:
  bonds, 171–172
  expenditure, 152, 153, 156, 163, 164, 165–166, 167, 168, 169, 170, 171, 185
  intervention, 49–50, 66–84,163–179, 226–233, 234–253
  supply, 148
Grants, 184
Green belt, 54, 189, 192, 200–201
Green wedges, 202
Grey areas, 204
Gross annual returns, 218–221
Growth, 157–158, 162

Hazardous materials, 67
Health and safety regulations, 21–22, 57, 70–72, 182, 220, 222
Historic preservation orders, 217
Historical buildings, 225, 237, 252
Home improvement grants, 70
Homelessness, 50
Homogeneity, 107–108
Horizontal integration, 118
Horizontal summation, 17, 23
Hot money, 160
House:
  builders, 55, 85
  price: earnings ratio, 48
Housing:
  birth rate, 49,
  boom, 55–57, 64, 145, 157, 166, 174
  construction time lags, 46–47
  decentralization, 198–199
  demand, 46, 47–48
  demography, 49, 174, 199
  divorce rate, 49
  economic life, 223
  elderly, 49
  equity, 52
  externalities, 240–241, 243, 244
  fixity of location, 52
  gentrification, 199–200
  government policy, 49–50
  imperfect competition, 51
  income, 47–48
  inner-city, 223
  interest rates, 176
  investment good, 45, 47
  labour mobility, 50
  land, 46, 47, 55–57
  land-use, 191, 193, 194
  lending institutions, 50–51
  market, 45–51
  migration, 49
  owner-occupied market, 45–52
  price, 46–47,50
  productivity, 47
  recession, 63–64, 109–111, 164
  renovation, 199
  stock, 47
  subsidies, 165
  substitutes, 48
  supply, 46–47
  tastes, 48

Illegal market, 76
Imperfect competition, 104, 113–121
Imperfect information, 249
Imports, 147, 149, 153, 155, 156, 160
Improvement grants, 165, 224, 242
Income:
  distribution, 145
  elasticity, 41–43
Industrial parks, 72, 243
Industrial property, 144–146, 150, 152, 154, 157, 160, 164, 166, 174, 191, 193, 194, 196–198, 202–203, 207, 222, 224
Industrial structure, 101–119
Industrialization, 195
Inequity, 25, 158
Inelasticity, 35–37, 39, 46
  demand, 35–36, 39
  land, 52–53
  supply, 35–37
Inference approach of valuation, 230
Inferior goods, 10–12, 42
Inflation, 137, 145, 151, 157, 158–159, 164, 167, 175, 177, 211
Inflationary gap, 167, 173
Information technology, 197
Injections, 151, 152, 156
Inner-city decline, 202–204
Inner-urban areas, 196–197, 198
Innovation, 130–135
Input prices, 22–23, 66
Interdependency, 235
Interest rates, 111, 144, 171, 172, 173, 174, 175, 176, 225
Internal costs, 227
Internal rate of return, 215–216
Investment, 158, 159, 162, 168, 171, 176, 177, 185
Investment appraisal, 206–233
  average rate of return, 211–212
  conventional techniques, 208, 209–213, 216–217
  discounting techniques, 208, 213–217
  internal rate of return, 211–216
  macro-level, 207–208
  micro-level, 206–207
  net present value, 213–215
  pay-back, 209–211
  private sector, 206–225
  public sector, 226–233

Keynes, J.M., 178–179

Labour, 79–81, 148
Labour productivity, 94, 181
Land:
  banking, 56
  building land, 52–57
  general, 19–20, 52–57
  housing land, 46–47
  inelasticity of supply, 52, 196
  intensive use, 54–55
  market, 52–56
  sub-division, 53–54
  use controls, 72–73
  use models, 191
  use planning, 72–73, 234, 235, 243, 249, 250–251, 252, 253
Law of diminishing returns, 93, 240
Law of variable proportions, 93, 240
Leakages, 151, 153, 156, 170
Lesser developed countries, 136
Life-cycle hypothesis, 179
Listing, 225
Local government, 185
Long-run planning curve, 97
Loss, 109–110, 113–114
Luxury/necessity syndrome, 38–39.

Macroeconomics, 141
Macroeconomic models, 143–156
Management, 100–101, 129, 132–134, 136
Managerial economies, 100
Marginal cost, 92–93, 94–95, 96
Marginal private benefit, 237–241
Marginal propensity to consume, 144, 152, 154, 155, 156, 166, 169, 174–175, 238–241
Marginal propensity to save, 144, 174, 175
Marginal revenue, 89, 96
Marginal social benefit, 241
Marginal social cost, 238–241
Market:
  analysis, 3–82
  equilibrium, 23–26
  failure, 1, 23, 50, 234–253
    externalities, 234–246
    imperfect information, 249–250
    monopolies, 247–249
    public goods, 246–247
  forces, 23–26
  price, 23–28
  system, 1
  theory, 3–29
Materials supply sector, 113–121, 152, 165
Maximization of social welfare, 86
Maximizing sales, 85
Maximum price controls, 73–79
Mergers, 118, 136
Merit goods, 252
Methodology, xi–xv

Migration, 145
Minimum price controls, 79–81
Minimum wage legislation, 80–81
Mixed economy, 1, 143
Modern management, 132–134
Modernization, 207
Modigliani, F. 179
Monetary policy, 49–50, 163, 173–179
  contractionary, 175–176
  expansionary, 174–175
  limitations, 176–179
Monocentric, 189
Monopolies and mergers commission, 118
Monopolistic competition, 104
Monopoly, 104, 113–121
  power, 115–116
Moonlighting, 126
Mortgage(s), 48, 164, 174, 175, 176–177
  insurance, 164
  interest tax relief, 49, 77
Multi-centered urban areas, 189
Multinational corporations, 136, 138
Multiple occupancy, 194, 198, 207, 223
Multiplier process, 144, 145, 146, 151–156, 157, 159, 160, 161, 165, 166, 167,169, 170, 171, 174, 175, 185, 227
  negative multiplier, 151, 153–155, 159, 160, 161, 166, 167, 170, 174, 175
  positive multiplier, 151, 152–153, 165, 166, 171, 175, 185, 227

Nationalization, 79
Necessities, 39
Negative externalities, 235–240, 241–245, 246
Negative multiplier, 151, 153–155, 159, 160, 161, 167, 170, 174, 175
Net annual returns, 219–221
Net present value, 213–215
Net social benefit, 227–228
Newly industrialized countries, 135
Non-excludability, 246
Non-pricing, 235
Non-rivalry, 246
Normal goods, 10–12, 42
Normal profit, 95, 108–110
Normative statements, xiv

Obsolescence, 196, 210
Office buildings, 157, 164, 166, 190, 191, 200, 207, 245, 250
Oil market, 135
Oligopoly, 104
Open economy, 143
Open public space, 246
Operating costs, 206, 218–221
Opportunity cost, 63, 95, 181, 210, 212, 230, 238
Overheads, 91
Overheating, 162, 167, 173, 175
Overseas construction, 135–138, 161, 176

Payback period method, 209–210
Perfect competition, 104, 105–113
Perfect elasticity, 34
Perfect inelasticity, 37
Perfect knowledge, 111
Periodic rent reviews, 219
Permanent income hypothesis, 11, 179
Personal loans, 176
Planning *see* Land use planning
Planning constraints, 54
Plot density, 88
Point elasticity, 29–30
Policy failure, 168
Pollution, 235–237, 240, 243, 244
  air, 235–236, 239, 240, 245
  noise, 236
  water, 236, 245
Positive externalities *see* Externalities
Positive multiplier, 151, 152, 153, 165, 166, 171, 175, 185
Positive statements, xiv
Present value, 213
Previous income, 179
Price:
  ceilings, 73–79
  control policies, 73–81
  elasticity,
    demand, 29–41
    supply, 31–41
  floors, 79–81
  takers, 85, 106
Primary data, xii
Private costs, 227
Private goods, 246
Producer:
  subsidies, 69, 79
  taxes, 66–68, 78
Productivity, 61, 110, 124, 125–135, 224
Profit:
  abnormal *see* Supernormal profit
  maximization, 85, 95–96, 105–113, 113–120
  maximizing rule, 95–96
  normal, 95, 108–110
  sharing, 183–184
  supernormal, 108–109, 113, 115, 118, 119, 120, 206
Propensity to consume, 144
Propensity to import, 145, 146

Public goods, 246–247
Public pressure groups, 217
Public sector, 146–147, 165–166, 226–233
Pump-priming, 171

Questionnaire technique of valuation, 230

Ransom price, 248
Ratchet effect, 179
Rationing, 79
Raw materials, 98
Recession, 150, 162, 164, 173, 174
Recovery, 162
Redevelopment, 207, 217–225
Red-lining, 50–51
Refurbishment, 194, 207, 218, 221
Regional variation, 150–151
Relative income hypothesis, 179
Relative price elasticity elasticities, 33–41
Relocation packages, 183
Renovation grants, 49, 70, 199
Rent:
  ceiling, 75–79
  rebate, 77
Repair and maintenance, 76, 105, 152, 157, 164, 165, 176
Repossession, 164
Research:
  and development, 99, 119
  economies, 99
Residential property *see* Housing
Residual, 55–57
Restrictions on financial institutions, 176, 178
Retail parks, 174
Retail price index, 158
Retail property, 144–146, 152, 154, 157, 164, 166, 174, 190, 194, 200, 202–203, 280
Revenue, 35, 86–90
  average revenue, 90, 106, 107
  marginal revenue, 89–90, 106, 107
  total revenue, 86–90
Ribbon development, 200
Risk, 208, 219, 231–232
  aversion, 208, 215
  pooling, 232
  premium, 215

Sales revenue maximization, 85, 120
Savings, 153, 155
Secondary data, xii
Security of tenure, 75, 77
Shadow pricing, 229–230
Skilled labour, 101
Slums, 195, 198
Smith, Adam, 133
Social benefits, 227–228
Social costs, 72, 164, 202, 227–228
Social rate of time preference, 231
Social welfare, 226
Specialization of labour, 133
Speculative builders, 41, 46, 47
Spill over costs, 227, 235
Standard industrial classification, 122
Steady-state, 157, 162, 163
Stock, 47, 52
Structural unemployment, 182
Sub-contractors, 124, 125, 128, 130, 134
Subsidization:
  employers, 81, 184
  householders, 185
  landlords, 76–77
  producers, 69, 136
Substitutability, 38–39, 43–44
Substitutes, 12,
  in demand, 12–14
  in supply, 21, 48, 58–59
Suburbia, 198
Supernormal profit, 108–109, 113, 115, 118, 119, 120, 206
Supply:
  consumers, 148
  determinants of, 18–23
  excess, 25
  expectations, 20–21
  firms, 148
  function, 18
  government, 148
  housing, 46–47
  input prices, 22, 66–68
  labour, 148
  management, 181
  of land, 19–20, 52–53
  price, 18–20
  price elasticity, 29–38, 40–41
  relative elasticity, 33–35, 58–59, 149–150
  relative inelasticity, 36–37, 58–59, 149–151
  shortages, 27–28
  substitutes, 21, 48, 58–59
  technological change, 21–22, 70–71
  time-lags, 46–47
  total market supply, 23
Supply-side economics, 151, 179–184
Supply-side policies, 150, 163, 181
  grants and subsidies, 184
  income tax, 181–182
  instruments, 181

profit-sharing, 183–184
trade unions, 182
unemployment, 182–183

Tastes, 15–16, 48, 58–59
Taxation, 66, 144, 153, 154, 156, 163, 164, 166, 167–168, 170, 173, 181–182, 184, 185
brackets, 170, 173
consumers, 68
energy, 67
holiday, 185
pollution, 67–68
producer, 66–68, 78
safety, 67
Taylorism, 133–134
Technological change, 21–22, 70–71, 97, 131–132
Test discount rate, 231
Testing, xiv
Theory, xii–xiv
of the firm, 103–121
Time-lags, 41, 47, 166, 168
Time and motion studies, 133
Time-value of money, 209, 212
Tolls, 239
Total costs, 85, 92, 93–94, 227
Total revenue, 85, 86–89
Trade unions, 182
Traffic congestion, 236, 237–240
Transfer earnings, 62–63
Transfer pricing, 136

Uncertainty, 169–170, 208, 211, 232
Unemployment, 80–81, 157, 159–160, 164, 182–183, 203
benefits, 173–183
University accommodation, 5–28
Urban areas, 189
Urban economic policy, 234
Urban economy, 190
Urban sprawl, 200–202
Urban structures, 190
Urbanization, 194–196, 203, 222
pull factors, 195
push factors, 196

Vacancies, 151, 158, 164
Variable costs, 91–92, 93–94, 227
Vertical integration, 118
'Very long-run', 97
Vibrational damage to buildings, 236–237

Wages of building labour, 62
Wage-floor, 80–81,
Willingness to pay, 230

Zones of transition, 194
Zoning, 243